Lignocellulose Pretreatment for Improved Bioenergy Recovery

Lignocellulosic biomass is a promising feedstock for second-generation bioenergy due to its low value and abundance. Different types of lignocellulosic biomass require specific pretreatment methods to enhance biorefinery processes. This book provides state-of-the-art information on recent developments in available pretreatment methods. It offers an overview of structural analysis of pretreated lignocellulosic biomass using various techniques, including optimization and prediction of bioenergy production with the help of statistical and artificial intelligence models.

Features:

- Focuses on lignocellulosic biomass pretreatment for bioenergy production
- Discusses environmental and social impacts, as well as economic approaches, of different pretreatment methods
- Covers the economic realities and environmental implications of each pretreatment technique
- Evaluates the performance of individual pretreatment techniques on biogas and methane yields
- Includes applications of statistical, artificial intelligence, and kinetic models

This book is intended for graduate students and researchers in bioprocessing, bioresource management, bioeconomy, and sustainability.

Lignocellulose Pretreatment for Improved Bioenergy Recovery

Kehinde O. Olatunji and Daniel M. Madyira

CRC Press
Taylor & Francis Group
Boca Raton London New York

CRC Press is an imprint of the
Taylor & Francis Group, an **informa** business

Designed cover image: © Bahman Fazeli-Nasab

First edition published 2027
by CRC Press
2385 NW Executive Center Drive, Suite 320, Boca Raton FL 33431

and by CRC Press
4 Park Square, Milton Park, Abingdon, Oxon, OX14 4RN

CRC Press is an imprint of Taylor & Francis Group, LLC

ISBN: 978-1-041-00461-5 (hbk)
ISBN: 978-1-041-06109-0 (pbk)
ISBN: 978-1-003-63381-5 (ebk)

DOI: 10.1201/9781003633815

Typeset in Times
by Apex CoVantage, LLC

Contents

PART II Lignocellulose biomass pretreatment

Authors' biography

Kehinde O. Olatunji received his PhD in mechanical engineering from the University of Johannesburg, South Africa, in 2023 and a Master of Technology in Agricultural Engineering from Ladoke Akintola University of Technology, Ogbomoso, Nigeria, in 2019. He is a postdoctoral research fellow at the University of Johannesburg, South Africa, and a research coordinator at Process Energy and Environmental Technology Station (PEETS), University of Johannesburg, South Africa. His research interests are waste-to-energy, biorefinery, waste management, and the application of artificial intelligence to energy recovery from organic wastes. He has authored and co-authored several peer-reviewed articles in reputable international journals and has presented at various academic/renewable energy/waste management conferences. He is currently co-supervising master's students researching energy recovery from biomass and working on applying the findings with PEETS. Kehinde is a corporate member of the Nigerian Institution of Agricultural Engineers (NIAE) and the Nigerian Society of Engineers (NSE) and a registered engineer with the Council for Regulation of Engineering Practices in Nigeria (COREN). He holds an academic excellence award as PhD student in the Faculty of Engineering and the Built Environment of the University of Johannesburg in 2022.

Daniel M. Madyira is a registered professional engineer and currently an associate professor of mechanical engineering in the Department of Mechanical Engineering Science in the Faculty of Engineering and the Built Environment at the University of Johannesburg. He has taught a wide range of core mechanical engineering courses, including machine design, fluid dynamics, thermodynamics, strength of materials, and biomass processing. He is a highly experienced mechanical engineer and academic with over 30 years of experience in both academia and industry. He is passionate about engineering design with special expertise in machine component stress analysis and thermal system analysis. He has developed several mechanical designs addressing a range of contemporary societal problems, including biomass briquetting and solar crop drying, as well as advanced mechanical systems such as rotating bending fatigue testing machines. His research interests range from high-speed machining of titanium and fatigue of titanium and composites to natural composites, biomass briquetting, and biomass combustion modeling, as well as the fracture behavior of materials produced using modern manufacturing techniques such as additive manufacturing. Additionally, he focuses on post-weld heat treatment of steels using weld beads. He is also involved in several industry-related activities, addressing specific problems within the industry. This makes his academic work relevant to industry and fruitful to his students. He has edited two conference proceedings, published two chapters of a book, several journal papers, and more than 150 refereed conference papers. He has registered one patent and graduated seven PhD students as supervisor and co-supervisor, as well as 34 M.Sc. and M.Phil. students, M.Tech. students, and supervised more than 20 international M.Sc. interns. He is currently supervising and co-supervising 10 PhD students and more than 20 master's students.

Preface

The rise in global demand for fossil fuels and their widespread consumption have resulted in energy shortages and environmental degradation, which have become a major challenge. With the expected further increase in global energy consumption over the next decade due to economic development, population growth, and social pressure. The need for energy security and environmental sustainability has necessitated the need for the search for alternative energy sources that can replace fossil fuels. Bioenergy from biomass has been identified as an alternative source of energy that can be utilized for heating, cooking, cooling, and electricity generation and can also be purified and injected into the grid. Bioenergy can be generated from biomass, which is a waste from other activities. Lignocellulose biomass from agricultural residues, forestry residues, algae, energy crops, wastewater sludge, and the organic fraction of municipal solid waste are the most available and economically viable sources of feedstock for bioenergy production.

Lignocellulose biomass is a complex structure with lignin, cellulose, and hemicellulose intertwined together. Lignin provides strength to the biomass, hemicellulose binds the component, and cellulose, which is composed of glucose, provides structural support for the biomass. Hence, the conversion of biomass into bioenergy is typically inefficient and labor-intensive. Therefore, the application of pretreatment is required to improve the amorphous content, increase the porosity of the porous matrix, and loosen the cellulose from hemicellulose and lignin. Different pretreatment methods include chemical, biological, thermal, mechanical/physical, and nanoparticle additives, as well as combined methods. Some of these methods can improve bioenergy production but can also generate inhibitory/toxic compounds, such as furfural, 5-methylfurfural, and acetic acid, which have a negative impact on bioenergy production. Various pretreatment techniques produce varying degree of impact; therefore, investigating different pretreatment techniques directly through the test data is not accurate. Therefore, a scientific, economically viable, and highly efficient pretreatment technique is required based on the evaluation standard for an efficient biorefinery of biomass for bioenergy production. This book explores the details of pretreatment methods to establish reasonable pretreatment models that can be used to optimize process conditions. Optimum methodological conditions are required, and a matching method is required to address the common limitations of lignocellulose biomass pretreatment. This book presents the application of pretreatment techniques to lignocellulose biomass, highlighting their merits and drawbacks, and proposes some future work required in the field for successful bioenergy production, both at laboratory and commercial scales.

Acknowledgment

The authors acknowledge the University of Johannesburg, Johannesburg, South Africa, for its support of this study.

Introduction

The search for alternative and sustainable energy to replace fossil fuels has become a hot topic in the research community, among industrialists, and the government. Several studies have proposed utilizing lignocellulose biomass for bioenergy production in recent years. Lignocellulose biomass, as a promising feedstock for second-generation bioenergy from lower-value and abundant sources, holds great promise to not only mitigate the challenges of excessive fossil fuel burning and serve as replacements for it but also significantly improve its ecological impact globally. However, some of this lignocellulose ends up as waste, polluting the environment, or is burned directly, resulting in the loss of its inherent energy. This is due to their composition, which includes phenolaldehyde polymer lignin, polysaccharide cellulose, hemicellulose, and other soluble polar and non-polar components, which have hindered the effectiveness of bioenergy technologies. To overcome this challenge, the application of pretreatments is crucial.

Different types of lignocellulose biomass require specific pretreatment methods to enhance biorefinery downstream technology. As one of the most crucial components of developing biomass-based processes, biomass pretreatment requires broad information, fundamental data, and an understanding of one of the most promising renewable energy sources. This book provides state-of-the-art information on recent developments in available pretreatment methods. It provides evidence-based scientific information on advanced and innovative pretreatment techniques for lignocellulose biomass, facilitating further processing. Feedstock pretreatment has been identified as a costly step in biomass biorefineries. With the significant progress made in developing lignocellulose biomass-based biorefineries, technically and financially viable pretreatment technologies are increasingly gaining global focus.

This book offers a comprehensive overview of recent advancements in methods for lignocellulose biomass pretreatment. It is divided into three sections, comprising eleven chapters, with each section focusing on an important study in bioenergy. The first section of the book focuses on biomass, biorefinery, and bioenergy. The subject of biomass classes, sources, generation, and their conversion technology into different bioenergy products was discussed under two chapters. The second section provides a detailed description of lignocellulose biomass pretreatment techniques, including their merits, limitations, and environmental impacts. This section comprises six chapters that discuss various categories of pretreatment methods, along with examples of how they impact bioenergy yield. A chapter in this section discusses the techno-economic analysis of pretreatment methods and examines the factors to be considered before a pretreatment method is deemed economically viable. The last three chapters in the third section discussed the impacts of pretreatment methods on the morphological, functional, and crystallinity index of the biomass. The section also discusses the merits, limitations, and future perspectives of biomass pretreatment for an improved biorefinery process. The last chapter of the book overviews the

DOI: 10.1201/9781003633815-1

application of various statistical and artificial intelligence models where pretreatment is considered as input process parameters. It considered the use of models to fit and predict bioenergy production from pretreated lignocellulose biomass, as well as their merits and limitations.

This book addresses some of the challenges facing researchers in environmental, mechanical, agricultural, chemical, and materials engineering, as well as biotechnologists, biochemists, biologists, and technology professionals seeking affordable and sustainable solutions for renewable energy production and waste management. It is an important book for policymakers and education professionals who teach renewable energy and waste management.

PART I

Biomass, biorefinery, and bioenergy

1 Biomass

1.1 INTRODUCTION

Biomass is a bright, renewable, and abundant material that can be utilized to produce bioenergy, addressing the challenges of energy security and environmental degradation. It is an effective carbon management material for carbon neutrality, with significant capacity for recyclability, natural availability, sustainability, and biorenewability [1]. Biomass is a composition of all living and organic materials that are classified according to their generation and sources. Economic, cultural, technological, and social parameters can affect the optimum technical applications of different classes of biomass available in varying components of the environment. Biomass consists of living materials and is grouped into lignocellulose and non-lignocellulose biomass [2]. Lignocellulose biomass is a composition of lignin, hemicellulose, cellulose, various extractives, and ash [3,4]. This includes agricultural residues, forest residues, municipal waste, and industrial waste. However, in addition to cellulose and lignin, macro- and microalgae have a higher quantity of saccharides, lipids, proteins, and inorganics [5]. It is envisaged that biomass will produce around two-thirds of the entire renewable energy that will be used by 2050 [6]. These large amounts of sustainable carbon materials are the most promising substrate to produce bioenergy [7]. Recent technological advancements have enabled the use of various biomasses as feedstock in biorefineries to support the energy transition and sustainable waste management.

1.2 BIOMASS AND ITS SOURCES

Biomass is the organic material present in the biosphere; it can be from animal or plant origins, and the ones that are produced from their natural or artificial conversion [8]. It is readily available globally and can be sourced from agricultural residues, domestic waste, and industrial by-products, representing a developing renewable energy source with considerable potential [9,10]. Biomass can be burned directly in waste-to-energy conversion plants to generate energy, making it a useful energy source. It has been utilized in boilers to release heat for domestic and industrial purposes [11]. It is worth noting that the direct combustion of biomass is not always feasible using existing equipment, and most of the time, it requires physicochemical analysis or some form of treatment to meet the standards of conventional fuels. Compared to fossil fuels, biomass district heating (BDH) achieves a 100% reduction in CO_2 emissions while increasing energy efficiency because biofuels are less expensive. BDH is a very effective system for integrating natural energy resources within urban environments. Biomass is available in various sources and forms, such as seed waste, paper waste, agricultural residues, wood, domestic waste, sawdust, municipal waste, wastewater, etc. [12–14]. Some of these biomass's physicochemical properties directly support their use as fuels, while others need varying degrees

DOI: 10.1201/9781003633815-3

of pretreatments using various technologies before being used as fuel. Biomass as a renewable energy source has its merits and limitations. Biomass sourced from the forest helps clean the forest and prevent forest fires within the forest's territory. It provides job opportunities by continuously gathering raw materials for energy production and processing. Energy generated from biomass contains lower carbon content and supports a net-zero emission approach. Biofuel production from biomass has recently become a significant interest in the scientific community [15]. One of the major limitations of biomass processing is the microstructural arrangement, which lowers the efficiency of production technologies [16].

1.3 GENERATIONS OF BIOMASS

Biomass for biofuels is categorized into four generations based on the raw materials and conversion technologies used. The influence of using this biomass as biofuel feedstocks on food for humans and animals is also a major criterion in categorizing these feedstocks into generations. They are the first, second, third, and fourth generations of biomass for biofuel production.

1.3.1 First generation

The first-generation biomass refers to biofuel feedstocks originating from edible crops, such as sugarcane, rapeseed, potato, cassava, sunflower, whey, oil palm, cereals, barley, and soybeans, grown on arable soils. They are grown through well-organized traditional methods, and the major biofuels from first-generation biomass are bioethanol and biodiesel [17]. These biomasses usually consist of oil, starch, or sugar that can be converted to final products through fermentation by yeast or transesterification [18]. The technology for converting these biomasses has reached the most advanced levels and is available on a commercial scale. It has been observed that since 2000, about 90% of the cumulative feedstocks for biofuel investment in tropical countries are from first-generation feedstocks [19]. First-generation biomass requires a large area of cultivable land, along with other resource inputs, which contrasts with the world's food production. Another demerit of these feedstocks in bioenergy production is their competition with the food supply, which causes fluctuations in food prices. For instance, the production of bioethanol from sugarcane will significantly influence the demand and price of sugar production. Also, bioenergy from these biomasses depends on the price of those products from the international market, which will affect the cost of production [20].

1.3.2 Second generation

Second-generation biomass was identified as a means to overcome the challenges associated with first-generation biomass as a biofuel feedstock. This biomass originates from various sources, including both edible and non-edible sources, plants, and waste materials. The major advantages of this biomass over first-generation biomass are its abundance, low cost, and noncompetitive with the food supply [21]. This biomass generation uses resources judiciously and encourages sustainable waste management [22]. These biomasses are mostly lignocellulose in nature and are categorized into three categories: quasi-homogeneous (e.g., forest and agricultural residues), homogeneous (e.g., wood chips), and non-homogeneous (e.g., wastewater sludge, municipal solid waste (MSW)) [19,23]. The two major conversion technologies for biomass generation are biological and thermochemical conversion. Aside from having good potential for biofuel production, it can also produce high-value chemicals and adhesives due to the presence of aromatic monomers in its composition. It has been observed that this second-generation biomass can serve as new raw materials for bio-adhesives and bioplastics. Recently, it has been reported that the construction industry has identified lignin as an alternative to bitumen asphalt binder. The bio-bitumen derived from lignin-based materials can help minimize the impact of asphalt on global warming [24]. However, second-generation biomass is complex and challenging to convert into bioenergy, requiring the application of pretreatment techniques for efficient processing [25]. They are lignocellulose in nature, comprising lignin, hemicellulose, and cellulose, which are strongly linked. Cellulose is a long-chain polysaccharide that consists of 6-carbon glucose monomers, while hemicellulose is a short-chain branched polymer with an amorphous structure and has 5-carbon monomers like xylose. Hemicellulose and cellulose are strongly intertwined to form a fiber that lignin fills in to strengthen the plant cell wall. The lignin portion is a complex polymer composed of no sugars but aromatic alcohols [26]. Lignin is the outer layer covering hemicellulose and cellulose, and it resists hydrolysis unless pretreatment is applied [27]. Therefore, pretreatment is necessary before converting these feedstocks into bioenergy. The essence of pretreatment is to depolymerize the lignin and expose cellulose and hemicellulose for bioconversion. Chemical, biological, mechanical, thermal, and nano-additive treatments can be used [25]. The lignin removed can be used to produce chemicals and some organic acids [28]. Pretreatment applications have different impacts on the microstructural arrangement, functional group, and crystallinity of the biomass. Chemical pretreatment degrades the structural arrangement of biomass through the swelling of cells, whereas biological pretreatment utilizes the power of microorganisms to disrupt the arrangement. Mechanical pretreatment uses different mechanical means to produce a new surface area that enhances biorefinery. The alteration of biomass at higher temperatures to break down its rigidity is thermal pretreatment, while nano-additive employs different nano-biocatalysts to influence bioenergy production [29,30].

1.3.3 Third Generation

Third-generation biomass is the feedstock from the microbial community. They are primarily algae and other aquatic biomass, which are replenished continually [21]. This biomass generation has been identified as having good potential to cater to future bioenergy needs and does not compete with the food supply or arable land. Third-generation biomass can be cultivated in different pathways, including open pond structures, modular closed photobioreactors, hybrid models, integrated systems, and heterotrophic fermentation. Different biofuels, such as methane, propanol, butanol, ethanol, gasoline, and biodiesel, can be produced from algae biomass [24]. Algae biomass has higher oil content than oil crops; some species contain more than

60–70% of the sample's dry weight, making them a potential candidate for biofuel production. Microalgae biomass generation is approximately 5–10 times higher than that of terrestrial plants [31]. Algae grow faster than other plants; it has higher photosynthetic strength and completes their growth cycle quickly [32]. Apart from their rapid growth, they need less land, which can be cultivated on non-arable lands. This category of biomass grows vast in oceans and other aquatic habitats [33]. Algae can capture carbon dioxide from the air for photosynthesis. Under the right conditions, it can capture up to 99% of carbon dioxide in the air, which can assist in reducing greenhouse gas emissions [34]. Due to this characteristic, there is a keen interest in cultivating algae as an affordable means of sequestering carbon. Like first- and second-generation feedstocks, this biomass category faces some limitations that hinder its upscaling in bioenergy production. Algae production on a large scale requires large volumes of water, which poses a significant challenge for countries that lack such facilities [17]. There is also concern about developing the proper technology process to remove lipids from algae, and some processes, such as dewatering and filtration, are required before extraction. At present, algae feedstock processing into biofuel is costly. Therefore, improved technologies are needed to produce a product that can effectively compete with fossil fuels in the market [17].

1.3.4 Fourth generation

Fourth-generation biomass is the extension of the third-generation bioenergy feedstock. These genetically engineered microorganisms, including yeast, cyanobacteria, microalgae, and fungi, convert carbon dioxide into fuel through photosynthesis [35]. It is a supplementary generation that produces a higher degree of energy than the energy contained in the input feedstock. Specific biomass that releases more energy than it does yields without putting more energy into the breakdown [21]. They are not usually viable agriculturally but have good potential for bioenergy release. This category of biomass is still at the development stage because it is a more advanced generation of biomass. Fourth-generation biomass theoretical concepts are divided into two: (1) The development of genetically modified plants that can absorb more carbon dioxide from the atmosphere, and (2) combining algae and oilseed plants produces a successful and better cross-genetic species [35]. This biomass generation plays a significant role in minimizing carbon dioxide release by capturing and storing it. It also helps reduce the release of other toxic gases and offers long-term economic benefits but requires a substantial investment [36]. This biomass generation can be converted to biofuel through genetic engineering, pyrolysis, upgrading, gasification, and solar-to-fuel routes. Some of the merits of this biomass generation include its high oil content, rapid growth, and low structural complexity [37]. Currently, very few works are ongoing in the conversion of this biomass, and the focus is mainly on microorganisms that have the ability to generate fuel directly from carbon dioxide. This process is believed to have been wholly developed by 2050 and has become a global breakthrough in the power sector [35].

It can be observed that each of these biomass generations has its merits and limitations, as presented in Table 1.1. First-generation technology is well-established globally, but it competes with food security and arable lands for other purposes.

TABLE 1.1
Merits and limitations of different biomass generations

Generations	Merits	Limitations
First	i. The technology is straightforward and well-established. ii. It is cost-competitive and scalable. iii. Intensive pretreatment is not required.	i. It competes with food security. ii. Inefficient use of land. iii. The yield is low, and available arable lands are limited. iv. Negative net energy
Second	i. They are non-food biomass and can be cultivated on marginal land. ii. Sustainable environmentally. iii. They are not controversial. iv. They release fewer greenhouse gases. v. It encourages a circular economy.	i. Upscaling to an industrial scale is a big challenge. ii. The investment cost is high. iii. Not available in every area. iv. It required extensive pretreatment.
Third	i. Biomass yield is high. ii. Required low energy input. iii. Broader availability. iv. Carbon neutral. v. It does not compete with food security.	i. Energy output is low. ii. The pilot scale is yet to be commercialized. iii. The downstream cost is high.
Fourth	i. Expected to be carbon neutral. ii. Yields higher because of genetic engineering. iii. Usually use syngas and carbon dioxide	i. Still at the conceptual/laboratory-scale phase. ii. Higher regulations are required because of safety issues.

These challenges have led to the search for second-generation biomass derived from agricultural and forestry residues, as well as municipal waste. However, these limitations are due to their complex structural arrangement, which requires advanced processing. Therefore, it is not easy to scale up to a large scale, and research is still ongoing to improve the efficiency of the technologies. Third-generation biomass from algae addresses the issues associated with first- and second-generation biomass by producing biomass rapidly and requiring less land. However, the technology to make it efficient and economical is still in its infancy and requires further investigation, investment, and development. Ultimately, the world must adopt more environmentally friendly methods to generate sustainable and renewable energy for human use. Because of this, biofuels are the fuel of the future, and their production might involve a mix of three generations of biomass rather than just one.

As presented in Table 1.2, it was observed that these generations of biomass have varying cost efficiencies and contributions to the greenhouse gas emissions. In terms of cost efficiency and CO_2 reduction, second-generation biofuels have better potential. This is because the feedstock is waste from other activities, requiring minimal cost and not producing extra CO_2 beyond the one released during initial production [38]. The first generation's cost efficiency is low due to its competition with the food supply. The cost efficiency of algae production is moderate due to the technology required, especially in regions with limited water resources. The processing

TABLE 1.2
Economy and environmental contribution of different biomass generations

Biomass	Generation	Bioenergy type	Conversion technology	Cost efficiency	CO_2 reduction potential	Ref.
Wheat straw	Second	Ethanol	Fermentation	High agricultural residue reduces feedstock cost.	High (80–90% reduction compared to gasoline)	[40]
Vegetable oil and animal fat	Second	Biodiesel	Transesterification	Moderate, depending on feedstock availability and cost.	High, above 78% compared to fossil diesel.	[41]
Sugarcane	First	Bioethanol	Fermentation	High due to low production cost in the tropical regions.	High, up to 90% compared to fossil fuels	[42]
Corn	First	Biodiesel	Transesterification	Moderate (50–70% due to land use emissions)	NA	[43]
Oil palm	First	Bioethanol	Fermentation	Low to medium, depending on the influence of deforestation cost and the region.	Low to moderate (20–50%) due to deforestation emissions.	[44]
Bagasse	Second	Biomethane	Anaerobic digestion	High, sugar industry waste lowers the price.	High, about 80–90% CO_2 reduction compared to gasoline.	[40]
Microalgae	Third	Biohydrogen	Dark fermentation	Medium, depending on the growth of algae.	It can reduce more than 500 tons of CO_2 /ha/year.	[45]

technology for third-generation feedstock is not yet mature, which affects cost efficiency; however, it is very efficient in CO_2 reduction [39].

1.4 CLASSIFICATION OF BIOMASS

Biomass resources are often referred to as biorenewable sources, derived from various organic sources, including plant matter from living organisms and animal waste. It can be generally classified into natural, energy crops, and residual biomass. Natural biomass is typically generated by nature without human intervention, and examples include forests, trees, and algae. Energy biomass is specifically cultivated to produce energy, and examples of this include corn, switchgrass, and energy-specific crops. On the other hand, residual biomass is released from different activities such as forestry waste (sawdust, wood chips, etc.), agricultural residues (shells, straw, bagasse, etc.), livestock waste (poultry waste, cow dung, etc.), and municipal solid waste (MSW) (food waste, yard waste, etc.). Furthermore, biomass is classified according to its origin as discussed below:

a. **Agricultural residues**
   Agricultural residues are the biomass that remains after harvesting and processing in agricultural activities. These materials are readily available and cheap, renewable feedstocks that can be utilized as a substitute for woody lignocellulosic biomass. This includes residues such as rice husks, groundnut shells, wheat straw, corn stalks, rice straw, cotton stalks, and sorghum stalks, among others [46–48]. It is one of the biomass sources available for bioenergy, which is the best way to utilize agricultural materials without compromising food, fiber, feed, and other agricultural products. These residues are left on the farm unused, where they degrade to add some soil nutrients or control soil erosion, while some are burnt openly and lose the inherent energy that can be converted into useful energy for the consumers [49]. It is an excellent feedstock for biorefineries, and its availability depends on the annual allocated land for different crops in an area. Agricultural residues have a similar composition, structure, and characteristics to those of other crops, making them suitable for bioenergy purposes. The reliance on agro-residues as a substitute for woody cellulosic fibers has increased recently because of the improved awareness of deforestation and environmental challenges of residue burning and due to the low cost compared to other crop fibers [50]. Additionally, the utilization of agro-residues for bioenergy production can reduce the reliance on wood resources, especially in countries where the forestry resources are limited.
b. **Forestry/woody biomass**
   Forest residues are the woody biomass that is left behind as waste after logging and other forestry activities, and this mostly includes cut or fallen limbs, stumps, treetops, and any other tree components that are not merchantable (i.e., malformed, diseased, and dead trees) [51]. These biomass residues from forestry are mostly burned for cooking, heating, or electricity generation. Residues from forestry can also be utilized to produce other

products, such as livestock bedding/litter, furniture, mulch, and wood chips. Gathering residues from forestry biomass is also useful for maintaining forest health, as it reduces wildfire risk and protects the forest from diseases and pests. During the processing of lumber, pulp, and other products, various waste and by-products, such as sawdust, bark, leaves, wood chips, and branches, are generated as wood processing residues [52]. Forestry residues are typically used to power mills, reducing reliance on fossil fuel-based energy and minimizing waste generation. These residues can also be used for briquettes and pellets for biofuel and applied as solid fuel [53].

c. **Livestock waste**
Livestock waste includes animal excrement, which is the transformation of other types of biomass (primarily grains, pasture, crop husk, and straw, etc.), and urine produced, and their mixture with beddings [54]. It can be produced from animals such as goats, cattle, horses, sheep, donkeys, and poultry. It contains energy, but the primary focus of biomass processing of livestock waste is to mitigate the challenges associated with their disposal rather than energy production. Stringent regulations on waste management, groundwater contamination, surface pollution, and odor control are driving the development of new manure management and disposal techniques. Biomass technologies offer promising methods for mitigating several environmental challenges associated with livestock waste. Anaerobic digestion, composting, and combustion are the primary technologies for managing biomass from livestock. Due to its composition, livestock waste is a vital feedstock for biogas production [55]. The quantity of livestock waste available varies by region and country due to differences in animal size and weight, as well as feeding practices. Therefore, the availability of livestock waste for bioenergy production depends on the region's environmental factors [56].

d. **Energy crop biomass**
Specific energy crops are a category of crops that are primarily grown to produce biomass for energy purposes. This category of crops is divided into two types: herbaceous and woody energy crops. Herbaceous energy biomass includes different perennial grasses like kochia, wheat grass, bamboo, sorghum, yellow thatching grass, etc., and other starchy and oil crops that are naturally utilized in the generation of bioethanol and biodiesel [57,58]. Woody biomass is fast-growing wood, such as silver maple, sweetgum, hybrid poplar, black walnut, and sycamore, that can be harvested within a short time and can be used to generate heat or produce electricity [59,60]. Although herbaceous biomass can be harvested faster than woody energy biomass, woody energy biomass offers better ecosystem benefits during its maturation period, as it helps enhance water quality, improve biodiversity, reduce erosion, and improve soil health. However, both types of biomass energy can be cultivated simultaneously through different agroforestry techniques, thereby diversifying farm income.

e. **Municipal solid waste**
MSW is the combination of waste produced from households, businesses, industrial, and construction activities, which are collected and managed

by the municipalities [61]. The rise in population and industrialization, improved standard of living, and consumer economy have led to an increase in the quantity and homogeneity of MSW released globally [62]. These wastes consist of food waste, yard trimmings, plastics, used paper, metals, textiles, glass, rubber, cardboard, wood, and wood products that can be separated into organic and non-organic waste [63]. Most of this waste ends up in landfills, contributing to environmental challenges. Sorted MSWs can be utilized as biomass for different waste-to-energy technologies to produce clean energy and chemicals. It can be converted into energy through various technologies, including thermal and biochemical processes. Waste-to-energy of MSW is a technology that converts MSW to heat or electricity during the treatment of MSW [63]. Incineration, biogas, biohydrogen, biodiesel, syngas, pyrolysis oil, and biodiesel are some of the energy sources that can be produced from MSW [64,65].

f. **Algae and aquatic biomass**
Biomass, such as macroalgae (i.e., seaweed), microalgae, and cyanobacteria, is a type of bioenergy feedstock derived from aquatic vegetation. Some of these microalgae store their energy as natural oils, which can be extracted and processed to release different biofuels, especially biodiesel and biogas [66]. Most of the macroalgae have bioactive compounds and exceptional carbohydrates that are extremely suitable for anaerobic digestion and fermentation to produce biogas and bioethanol, respectively [39]. Another important aquatic biomass for bioenergy production is cyanobacteria. It is also called blue-green algae and is an aquatic prokaryote that has photosynthetic pigments that help it convert solar energy to chemical energy through photosynthesis [67]. It is a varying group of prokaryotic biomasses that grows in various habitats ranging from saltwater to freshwater and extreme environments. They are characterized by special physicochemical properties that set them apart from other photosynthetic organisms. However, the commercialization of biofuel production of this class of biomass is still challenging due to their cytotoxicity, slow growth, energy-intensive processing, and low yield.

1.5 STRUCTURAL ARRANGEMENTS OF LIGNOCELLULOSE BIOMASS

Biomass for bioenergy production is primarily lignocellulose, a composition of aromatic polymer (lignin) and polysaccharides (cellulose and hemicellulose) [68], as illustrated in Figure 1.1. The three polymers are intertwined to form a strong and stable material. Lignocellulose biomass typically consists of 15–20% lignin, 40–50% cellulose, and 25–35% hemicellulose [60], which varies depending on the type, age, location, and other factors. Out of the three polymers, cellulose is a glucose-based polymer that is biodegradable but not soluble in water and some organic solvents. It is a structural component present in the cell wall and consists of d-glucose subunits bonded by β-1,4-glycosidic bonds, as shown in Figure 1.1. Cellulose has long polysaccharide links that are not branched and are parallelly arranged to produce

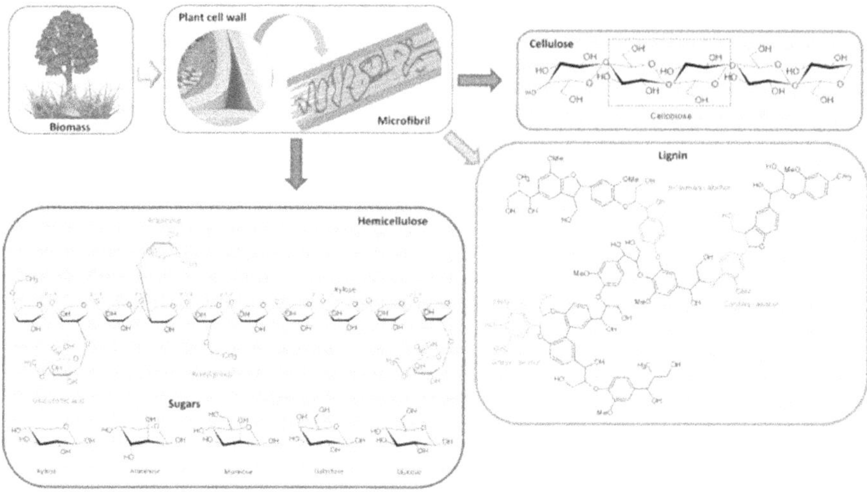

FIGURE 1.1 Structural components of lignocellulose biomass [76].

cellulose microfibrils, and is one of the most available natural biopolymers on earth. These microfibrils are strongly bound together by intra- and intermolecular hydrogen bonds to form a strong crystalline or amorphous arrangement. The cross-polarization-magic-angle spinning (CP-MAS) investigation reveals that the crystalline arrangement of cellulose consists of two types: Iβ and Iα [69]. It is primarily altered and solubilized using enzymatic hydrolysis or acid [70]. Another component of lignocellulose is the hemicellulose, which is a complex carbohydrate, a subdivided polymer with a heterogeneous composition of hexoses (galactose, mannose, glucose), pentoses (arabinose, xylose), and sugar acids (glucuronic, galacturonic, and 4-O-methyl-glucuronic acids). These sugars are bound by β-1,4-glycosidic and, at times, by β-1,3-glycosidic links [71] as illustrated in Figure 1.1. Glucomannan and xylan are the primary components of hemicellulose, which are easily degraded due to their hydrophilic nature and amorphous structure. It differs in composition across various biomass types. For instance, softwood is composed of glucans, arabinogalactans, xylans, and xyloglucans, whereas hardwood contains glucomannans and xylans [72]. Hemicellulose is strongly bound with cellulose microfibrils, pectin, and lignin to become a cross-bound network of heterogeneous materials consisting of hexoses and pentoses in the cell walls [73].

The third major component of lignocellulose biomass is lignin. It is a non-sugar-based polymer that has a three-dimensional network arrangement and is neither degradable nor soluble. It is a complex hetero-biopolymer composed of phenylpropane units linked together by aryl ether and C–C bonds. Lignin, alongside cellulose, is regarded as one of the most available biopolymers on earth [69]. Lignin is developed through oxidative reaction of three monolignols, which are trans-sinapyl, trans-p-coumaryl, and trans-coniferyl alcohols. During the formation of polymer from these monomers, the phenylpropane units are regarded as syringyl, guaiacyl, and p-hydroxyphenyl units [74,75]. The phenylpropanoid content of lignin is the primary challenge of degrading biomass due to its structural support, impermeability, and resistance against microbial attacks [76]. With its strong and resistant lignin, lignocellulose biomass is protected from biodigestion [25].

Lignocellulose biomass is a promising feedstock for producing reducing sugars, which are instrumental in the production of biofuels (e.g., bioethanol, biomethane) and other ancillary chemicals. However, the structural arrangement of lignocellulose biomass is complicated and limits the efficiency of bioenergy production. The cell walls of lignocellulose biomass are recalcitrant to biological and biochemical conversion, reducing the valorization efficiency. Lignin and hemicellulose are the main chemical components that limit access to cellulose content during conversion [77]. Therefore, the application of pretreatment is required to enhance the biodegradation of lignocellulose biomass. The aim of pretreatment is to alter the lignin protecting the cellulose, remove/redistribute hemicellulose, and reduce the cellulose crystallinity to make more oligosaccharides available for energy production [78].

1.6 APPLICATIONS OF LIGNOCELLULOSE BIOMASS

Biomass energy is a vital form of energy that has existed in various forms since ancient times. It has been applied for different purposes at both domestic and commercial levels as a renewable energy source and as an alternative to fossil fuels. The following are some of the common uses of biomass energy:

i. **Heat production:** This is the most common use of biomass energy that can be traced to human evolution. Heat can be generated in furnaces and boilers using biomass, such as woody biomass, wood pellets, agricultural residues, and dried livestock manure. This application can be utilized for building heating, space heating, water heating, and heating for greenhouses, hospitals, schools, and manufacturing equipment [79].
ii. **Electricity production:** Through gasification and combustion, biomass such as agricultural residues, wood residue, and energy crops can be used to generate electricity. This electricity can be integrated into the existing electricity grid to provide dispatchable energy or baseload power, thereby reducing dependence on fossil-fuel-powered energy sources.
iii. **Biofuel generation:** Different technologies such as fermentation, anaerobic digestion, thermochemical, and transesterification can be used to convert biomass into gaseous or liquid biofuels such as bioethanol, biogas, biohydrogen, and biodiesel. These biofuels can be used as transport fuel in ships, airplanes, and vehicles as renewable energy to substitute for diesel, jet fuel, and gasoline, thereby reducing greenhouse gas emissions in the transport sector [80,81].
iv. **Combined heat and power (CHP) generation:** Biomass can be used to generate heat and electricity simultaneously in a CHP system, improving the overall energy efficiency and minimizing the carbon footprint. The technique can produce heat and electricity for communities, district

heating networks, and industrial sectors as an economical and eco-benign source of energy.

v. **District heating and cooling:** Biomass can be used to generate heat or cold and supply it to multiple facilities or buildings within a district or community. In this case, biomass CHP equipment, geothermal heat pumps, or biomass boilers can be utilized for space and water heating, as well as air conditioning, in institutional, residential, and commercial buildings.

vi. **Cogeneration in the industry:** Biomass can be used to produce heat and power for industrial activities such as chemical production, pulp and paper production, pharmaceutical production, and food processing. In cogeneration, waste from industrial activities can be utilized as a source of energy within the plant, enhancing energy efficiency, reducing operating costs, and promoting sustainable waste management.

vii. **Off-grid applications:** Energy from biomass can be utilized to generate energy in remote or off-grid areas where there is little/no access to central energy facilities. This may include the application of biogas digesters or biomass gasifiers to power microgrids to produce energy for heating systems in the community [82].

1.7 CHALLENGES/LIMITATIONS OF LIGNOCELLULOSE BIOMASS APPLICATIONS

Biomass abundance, low cost, and readily available technology make it a promising energy feedstock with potential for other value-added products. Bioenergy derived from biomass is renewable, sustainable, and a promising candidate for enhancing energy security and environmental remediation. However, there are numerous challenges that hinder the full potential of biomass. Biomass is a heterogeneous, strong, and complex arrangement that makes it difficult to degrade into smaller particles during its conversion to bioenergy and other value-added products. Therefore, the efficiency of the biofuel generation technologies depends primarily on the physicochemical characteristics of the biomass, process optimization conditions, and pretreatment techniques [83]. The calorific value of biomass is relatively low, and a large quantity of biomass will be needed to produce the same amount of energy that can be generated from conventional hydrocarbons [84]. Biomass is characterized by a high percentage of oxygen, which will have a negative impact on its conversion to bioenergy. Therefore, oxygen must be removed before energy comparable to hydrocarbon fuel can be produced. This process will lead to additional resources, which may make the process uneconomical. Most of the available biomass consists of high moisture, which makes it unsuitable for certain conversion techniques, such as pyrolysis and gasification. The efficiency of biomass conversion is lower due to high moisture content, and fuels generated from this biomass can contain moisture. Also, part of the fuel generated from biomass with high moisture content will be required to vaporize the water in the fuel. Biomass bulk density is usually low (woody biomass 160–220 kg/m^3 and grass biomass 80–150 kg/m^3), which makes biomass handling a challenge and increases the transportation and storage expenses [85]. Government policies and institutional frameworks are key to the application of biomass. In many countries, the specific targets and goals for biomass energy development are not well-defined, which is hindering the progress of biomass energy. The development of biomass energy is slow in several countries due to insufficient funding for research and development, which has limited its acceptance because of the high cost and low level of technology. Furthermore, awareness of the application and benefits of biomass energy, particularly in terms of environmental protection, remains limited, while the knowledge and training among local people are inadequate [86]. To fully utilize the potential of biomass energy, these challenges must be addressed, and a more robust technology for biomass energy production and application must be developed.

1.8 CONCLUSION

Biomass is a renewable energy source that is generally available worldwide at a lower cost, as some of it is waste from other activities. There are four evolutions of biomass, and each generation has its merits and limitations. Most of this biomass is lignocellulose, which has a strong structural arrangement that hinders its conversion into energy. Energy released from biomass is renewable, with low greenhouse gas emissions during combustion, and is also sustainable for waste management. Renewable energy from biomass can be used for heating, cooking, cooling, CHP, and can also be fed into an existing grid for electricity generation. For a continuous supply of biomass for bioenergy, there is a need to select available, affordable, and competitive options that are compatible with the food supply and have lower energy requirements. Biomass must be efficiently utilized in each bioprocessing by ensuring that all biomass fractions are utilized to produce different products, either individually or in combination with other materials. Efficient use of biomass will support the circular bioeconomy while also helping to curb the climate change challenges arising from the burning of fossil fuels and improper waste management.

REFERENCES

[1] A.M. Elgarahy, A. Hammad, D.M. El-Sherif, M. Abouzid, M.S. Gaballah, K.Z. Elwakeel, Thermochemical conversion strategies of biomass to biofuels, techno-economic and bibliometric analysis: A conceptual review, J Environ Chem Eng 9 (2021) 106503. https://doi.org/10.1016/J.JECE.2021.106503.

[2] R.K. Srivastava, N.P. Shetti, K.R. Reddy, T.M. Aminabhavi, Sustainable energy from waste organic matters via efficient microbial processes, Science Total Environ 722 (2020) 137927. https://doi.org/10.1016/J.SCITOTENV.2020.137927.

[3] S. Jung, N.P. Shetti, K.R. Reddy, M.N. Nadagouda, Y.K. Park, T.M. Aminabhavi, E.E. Kwon, Synthesis of different biofuels from livestock waste materials and their potential as sustainable feedstocks—A review, Energy Convers Manag 236 (2021) 114038. https://doi.org/10.1016/J.ENCONMAN.2021.114038.

[4] M. Zborowska, H. Waliszewska, B. Waliszewska, S. Borysiak, J. Brozdowski, A. Stachowiak-Wencek, Conversion of carbohydrates in lignocellulosic biomass after chemical pretreatment, Energies 15 (2021) 254. https://doi.org/10.3390/EN15010254.

[5] D.C. Li, H. Jiang, The thermochemical conversion of non-lignocellulosic biomass to form biochar: A review on characterizations and mechanism elucidation, Bioresour Technol 246 (2017) 57–68. https://doi.org/10.1016/J.BIORTECH.2017.07.029.

[6] D. Gielen, F. Boshell, D. Saygin, M.D. Bazilian, N. Wagner, R. Gorini, The role of renewable energy in the global energy transformation, Energy Strategy Rev 24 (2019) 38–50. https://doi.org/10.1016/J.ESR.2019.01.006.
[7] Z.M.A. Bundhoo, R. Mohee, Ultrasound-assisted biological conversion of biomass and waste materials to biofuels: A review, Ultrason Sonochem 40 (2018) 298–313. https://doi.org/10.1016/J.ULTSONCH.2017.07.025.
[8] M.A. Perea-Moreno, E. Samerón-Manzano, A.J. Perea-Moreno, Biomass as renewable energy: Worldwide research trends, Sustainability 11 (2019) 863. https://doi.org/10.3390/SU11030863.
[9] F. Manzano Agugliaro, Gasification of greenhouse residues for obtaining electrical energy in the south of spain: Localization by gis, Interciencia 32 (2007) 131–136. http://ve.scielo.org/scielo.php?script=sci_arttext&pid=S0378-18442007000200012&lng=en&nrm=iso&tlng=es (accessed April 3, 2025).
[10] K.O. Olatunji, D.M. Madyira, Effect of acidic pretreatment on the microstructural arrangement and anaerobic digestion of Arachis hypogea shells; and process parameters optimization using response surface methodology, Heliyon 9 (2023) e15145. https://doi.org/10.1016/J.HELIYON.2023.E15145.
[11] M.A. Perea-Moreno, F. Manzano-Agugliaro, A.J. Perea-Moreno, Sustainable energy based on sunflower seed husk boiler for residential buildings, Sustainability 10 (2018) 3407. https://doi.org/10.3390/SU10103407.
[12] K.A. Adegoke, O.O. Adesina, O.A. Okon-Akan, O.R. Adegoke, A.B. Olabintan, O.A. Ajala, H. Olagoke, N.W. Maxakato, O.S. Bello, Sawdust-biomass based materials for sequestration of organic and inorganic pollutants and potential for engineering applications, Curr Res Green Sustain Chem 5 (2022) 100274. https://doi.org/10.1016/J.CRGSC.2022.100274.
[13] K.O. Olatunji, D.M. Madyira, Anaerobic co-digestion of alkali-pretreated groundnut shells and duck waste for methane yield optimization and sustainable environment, E3S Web Conf 433 (2023) 02005. https://doi.org/10.1051/E3SCONF/202343302005.
[14] M. Sillanpää, A. Khadir, K. Gurung, Resource recovery in municipal waste waters, Resour Recovery Municipal Waste Waters (2023) 1–399. https://doi.org/10.1016/C2021-0-01617-5.
[15] J. Popp, S. Kovács, J. Oláh, Z. Divéki, E. Balázs, Bioeconomy: Biomass and biomass-based energy supply and demand, N Biotechnol 60 (2021) 76–84. https://doi.org/10.1016/J.NBT.2020.10.004.
[16] C. Catherine, M. Twizerimana, Biogas production from thermochemically pretreated sweet potato root waste, Heliyon 8 (2022). https://doi.org/10.1016/j.heliyon.2022.e10376.
[17] S. Bardhan, S. Gupta, M.E. Gorman, A. Haider, Biorenewable chemicals: Feedstocks, technologies and the conflict with food production, Renew Sustain Energy Rev 51 (2015). https://doi.org/10.1016/j.rser.2015.06.013.
[18] B. Pedroli, B. Elbersen, P. Frederiksen, U. Grandin, R. Heikkilä, P.H. Krogh, Z. Izakovičová, A. Johansen, L. Meiresonne, J. Spijker, Is energy cropping in Europe compatible with biodiversity?—Opportunities and threats to biodiversity from land-based production of biomass for bioenergy purposes, Biomass Bioenergy 55 (2013) 73–86. https://doi.org/10.1016/J.BIOMBIOE.2012.09.054.
[19] M. Acheampong, F.C. Ertem, B. Kappler, P. Neubauer, In pursuit of Sustainable Development Goal (SDG) number 7: Will biofuels be reliable?, Renew Sustain Energy Rev 75 (2017) 927–937. https://doi.org/10.1016/J.RSER.2016.11.074.
[20] A. Maliha, B. Abu-Hijleh, A review on the current status and post-pandemic prospects of third-generation biofuels, Energy Syst 14 (2022). https://doi.org/10.1007/s12667-022-00514-7.

[21] N.B. Chanu, A.B. Singh, M. Chandrakumar Singh, Microwave assisted catalysis for biofuel production, Bioenergy Engineering: Fundamentals, Methods, Modelling, and Applications (2023) 427–445. https://doi.org/10.1016/B978-0-323-98363-1.00016-8.
[22] K.O. Olatunji, D.M. Madyira, O. Adeleke, Optimizing anaerobic co-digestion of Xyris capensis and duck waste using neuro-fuzzy model and response surface methodology, Fuel 354 (2023) 129334. https://doi.org/10.1016/J.FUEL.2023.129334.
[23] A. Sertolli, Z. Gabnai, P. Lengyel, A. Bai, Biomass potential and utilization in worldwide research trends A bibliometric analysis, Sustainability 14 (2022) 5515. https://doi.org/10.3390/SU14095515.
[24] S.J. Malode, S.A.M. Gaddi, P.J. Kamble, A.A. Nalwad, U.M. Muddapur, N.P. Shetti, Recent evolutionary trends in the production of biofuels, Mater. Sci. Energy Technol. 5 (2022) 262–277. https://doi.org/10.1016/J.MSET.2022.04.001.
[25] K.O. Olatunji, N.A. Ahmed, O. Ogunkunle, Optimization of biogas yield from lignocellulosic materials with different pretreatment methods: A review, Biotechnol Biofuels 14 (2021). https://doi.org/10.1186/s13068-021-02012-x.
[26] J. Zakzeski, P.C.A. Bruijnincx, A.L. Jongerius, B.M. Weckhuysen, The catalytic valorization of lignin for the production of renewable chemicals, Chem Rev 110 (2010) 3552–3599. https://doi.org/10.1021/CR900354U/ASSET/CR900354U.FP.PNG_V03.
[27] D.M. Madyira, K.O. Olatunji, Investigating the influence of different pretreatment methods on the morphological structure of Arachis hypogea shells, Mater Today Proc (2023). https://doi.org/10.1016/J.MATPR.2023.07.289.
[28] X.J. Ji, H. Huang, Z.K. Nie, L. Qu, Q. Xu, G.T. Tsao, Fuels and chemicals from hemicellulose sugars, Adv Biochem Eng Biotechnol 128 (2012) 199–224. https://doi.org/10.1007/10_2011_124.
[29] V.B. Agbor, N. Cicek, R. Sparling, A. Berlin, D.B. Levin, Biomass pretreatment: Fundamentals toward application, Biotechnol Adv 29 (2011) 675–685. https://doi.org/10.1016/j.biotechadv.2011.05.005.
[30] S.I. Mussatto, G.M. Dragone, Biomass pretreatment, biorefineries, and potential products for a bioeconomy development, Biomass Fractionation Technologies for a Lignocellulosic Feedstock Based Biorefinery (2016) 1–22. https://doi.org/10.1016/B978-0-12-802323-5.00001-3.
[31] Z. Sun, J. Liu, Z.G. Zhou, Algae for biofuels: An emerging feedstock, Handbook of Biofuels Production: Processes and Technologies. Second Edition (2016) 673–698. https://doi.org/10.1016/B978-0-08-100455-5.00022-9.
[32] R.H. Wijffels, M.J. Barbosa, M.H.M. Eppink, Microalgae for the production of bulk chemicals and biofuels, Biofuels Bioprod Biorefin 4 (2010) 287–295. https://doi.org/10.1002/BBB.215.
[33] K.O. Olatunji, D.M. Madyira, J.O. Amos, Sustainable enhancement of biogas and methane yield of macroalgae biomass using different pretreatment techniques: A mini-review, Energy Environ (2023). https://doi.org/10.1177/0958305X231193869/ASSET/IMAGES/LARGE/10.1177_0958305X231193869-FIG1.JPEG.
[34] X. Zeng, M.K. Danquah, X.D. Chen, Y. Lu, Microalgae bioengineering: From CO2 fixation to biofuel production, Renew Sustain Energy Rev 15 (2011) 3252–3260. https://doi.org/10.1016/J.RSER.2011.04.014.
[35] D. Gupta, S.K. Gaur, Carbon and biofuel footprinting of global production of biofuels, Biomass, Biopolymer-Based Materials, and Bioenergy: Construction, Biomedical, and Other Industrial Applications (2019) 449–481. https://doi.org/10.1016/B978-0-08-102426-3.00019-9.
[36] M. Rajamehala, R. Su, B. Gopalakrishnan, A. Muthu Kumara Pandian, M. Vijay Pradhap Singh, S. Chozhavendhan, Boundaries and openings of biorefineries towards

sustainable biofuel production, Biofuels and Bioenergy: A Techno-Economic Approach (2022) 3–22. https://doi.org/10.1016/B978-0-323-90040-9.00019-9.

[37] A.R.K. Gollakota, C.M. Shu, P.K. Sarangi, K.P. Shadangi, S. Rakshit, J.F. Kennedy, V.K. Gupta, M. Sharma, Catalytic hydrodeoxygenation of bio-oil and model compounds—Choice of catalysts, and mechanisms, Renew Sustain Energy Rev 187 (2023) 113700. https://doi.org/10.1016/J.RSER.2023.113700.

[38] K.O. Olatunji, D.M. Madyira, A.M. Kożuch, Biomass resources assessment and pretreatment technologies, Renewable Energy Technologies (2026) 3–31. https://doi.org/10.1016/B978-0-443-33771-0.00001-0.

[39] K.O. Olatunji, D.M. Madyira, J.O. Amos, Sustainable enhancement of biogas and methane yield of macroalgae biomass using different pretreatment techniques: A mini-review, Energy Environ 35 (2024) 1050–1088. https://doi.org/10.1177/0958305X231193869.

[40] D. Bunphan, P. Jaisil, J. Sanitchon, J.E. Knoll, W.F. Anderson, Estimation methods and parameter assessment for ethanol yields from total soluble solids of sweet sorghum, Ind Crops Prod 63 (2015) 349–356. https://doi.org/10.1016/J.INDCROP.2014.10.007.

[41] I. Nurfitri, G.P. Maniam, N. Hindryawati, M.M. Yusoff, S. Ganesan, Potential of feedstock and catalysts from waste in biodiesel preparation: A review, Energy Convers Manag 74 (2013) 395–402. https://doi.org/10.1016/J.ENCONMAN.2013.04.042.

[42] S. Barros, A. Specialist, This Report Contains Assessments of Commodity and Trade Issues Made by USDA Staff and Not Necessarily Statements of Official U.S. Government Policy Required Report-Public Distribution (n.d.).

[43] D. Neupane, Biofuels from renewable sources, a potential option for biodiesel production, Bioengineering 10 (2022) 29. https://doi.org/10.3390/BIOENGINEERING10010029.

[44] T. Srimachai, K. Nuithitikul, S. O-Thong, P. Kongjan, K. Panpong, Optimization and kinetic modeling of ethanol production from oil palm frond juice in batch fermentation, Energy Procedia 79 (2015) 111–118. https://doi.org/10.1016/J.EGYPRO.2015.11.490.

[45] M. Molazadeh, H. Ahmadzadeh, H.R. Pourianfar, S. Lyon, P.H. Rampelotto, The use of microalgae for coupling wastewater treatment with CO2 biofixation, Front Bioeng Biotechnol 7 (2019) 430241. https://doi.org/10.3389/FBIOE.2019.00042/BIBTEX.

[46] K.O. Olatunji, D.M. Madyira, N.A. Ahmed, O. Ogunkunle, Influence of alkali pretreatment on morphological structure and methane yield of Arachis hypogea shells, Biomass Convers Biorefin 14 (2024) 12143–12154. https://doi.org/10.1007/S13399-022-03271-W/METRICS.

[47] A.D. Olugbemide, A. Oberlintner, U. Novak, B. Likozar, Lignocellulosic corn stover biomass pre-treatment by Deep Eutectic Solvents (DES) for biomethane production process by bioresource anaerobic digestion, Sustainability 13 (2021) 10504. https://doi.org/10.3390/SU131910504.

[48] T. Suopajärvi, P. Ricci, V. Karvonen, G. Ottolina, H. Liimatainen, Acidic and alkaline deep eutectic solvents in delignification and nanofibrillation of corn stalk, wheat straw, and rapeseed stem residues, Ind Crops Prod 145 (2020) 111956. https://doi.org/10.1016/J.INDCROP.2019.111956.

[49] K. Sivabalan, S. Hassan, H. Ya, J. Pasupuleti, A review on the characteristic of biomass and classification of bioenergy through direct combustion and gasification as an alternative power supply, J Phys Conf Ser 1831 (2021) 012033. https://doi.org/10.1088/1742-6596/1831/1/012033.

[50] M. Röder, K. Chong, P. Thornley, The future of residue-based bioenergy for industrial use in Sub-Saharan Africa, Biomass Bioenergy 159 (2022) 106385. https://doi.org/10.1016/J.BIOMBIOE.2022.106385.

[51] J.A. Honorato-Salazar, J. Sadhukhan, Annual biomass variation of agriculture crops and forestry residues, and seasonality of crop residues for energy production in Mexico, Food Bioproducts Process 119 (2020) 1–19. https://doi.org/10.1016/J.FBP.2019.10.005.

[52] D.L. Grebner, P. Bettinger, J.P. Siry, K. Boston, Forest products, Introduction to Forestry and Natural Resources (2022) 101–129. https://doi.org/10.1016/B978-0-12-819002-9.00004-3.

[53] M.A. Waheed, O.A. Akogun, C.C. Enweremadu, An overview of torrefied bioresource briquettes: Quality-influencing parameters, enhancement through torrefaction and applications, Bioresour Bioprocess 9 (2022) 1–18. https://doi.org/10.1186/S40643-022-00608-1.

[54] A. Tolessa, Potential of biomethane-based energy production from livestock waste biomass resources in Ethiopia, Front Energy Res 11 (2023) 1249327. https://doi.org/10.3389/FENRG.2023.1249327/BIBTEX.

[55] P.A. Seglah, Y. Wang, H. Wang, C. Gao, Y. Bi, Sustainable biofuel production from animal manure and crop residues in Ghana, Energies 15 (2022) 5800. https://doi.org/10.3390/EN15165800.

[56] A. Tolessa, S. Zantsi, T.M. Louw, J.C. Greyling, N.J. Goosen, Estimation of biomass feedstock availability for anaerobic digestion in smallholder farming systems in South Africa, Biomass Bioenergy 142 (2020) 105798. https://doi.org/10.1016/J.BIOMBIOE.2020.105798.

[57] S.B. Ummalyma, D. Sahoo, A. Pudiyamadam, V.P. Adarsh, R.K. Sukumaran, T. Bhaskar, A. Parida, Sono-assisted alkali and dilute acid pretreatment of phragmites karka (Tall Reed Grass) to enhance enzymatic digestibility for bioethanol conversion, Front Energy Res 8 (2021) 594452. https://doi.org/10.3389/FENRG.2020.594452/BIBTEX.

[58] N.B. Khorasgani, A.B. Sengul, E. Asmatulu, Briquetting grass and tree leaf biomass for sustainable production of future fuels, Biomass Convers Biorefin 10 (2020) 915–924. https://doi.org/10.1007/S13399-019-00465-7/METRICS.

[59] J.C. Solarte-Toro, J.A. Arrieta-Escobar, B. Marche, C.A. Cardona Alzate, Effect of the lignin extraction process on the economics of a woody-based biorefinery, Computer Aided Chem Eng 50 (2021) 1871–1876. https://doi.org/10.1016/B978-0-323-88506-5.50290-4.

[60] P. Wang, B.H. Howard, Impact of thermal pretreatment temperatures on woody biomass chemical composition, physical properties and microstructure, Energies 11 (2017) 25. https://doi.org/10.3390/EN11010025.

[61] Q. Yang, L. Fu, X. Liu, M. Cheng, Evaluating the efficiency of municipal solid waste management in China, Int J Environ Res Public Health 15 (2018) 2448. https://doi.org/10.3390/IJERPH15112448.

[62] M.D. Meena, M.L. Dotaniya, B.L. Meena, P.K. Rai, R.S. Antil, H.S. Meena, L.K. Meena, C.K. Dotaniya, V.S. Meena, A. Ghosh, K.N. Meena, A.K. Singh, V.D. Meena, P.C. Moharana, S.K. Meena, Ch. Srinivasarao, A.L. Meena, S. Chatterjee, D.K. Meena, M. Prajapat, R.B. Meena, Municipal solid waste: Opportunities, challenges and management policies in India: A review, Waste Manag Bull 1 (2023) 4–18. https://doi.org/10.1016/J.WMB.2023.04.001.

[63] Z.Z. Rasmeni, D.M. Madyira, A review of the current municipal solid waste management practices in Johannesburg City Townships, Procedia Manuf 35 (2019) 1025–1031. https://doi.org/10.1016/J.PROMFG.2019.06.052.

[64] U. Di Matteo, B. Nastasi, A. Albo, D.A. Garcia, Energy contribution of OFMSW (Organic Fraction of Municipal Solid Waste) to energy-environmental sustainability in urban areas at small scale, Energies 10 (2017) 229. https://doi.org/10.3390/EN10020229.

[65] D. Moya, C. Aldás, D. Jaramillo, E. Játiva, P. Kaparaju, Waste-to-energy technologies: An opportunity of energy recovery from Municipal Solid Waste, using Quito—Ecuador as case study, Energy Procedia 134 (2017) 327–336. https://doi.org/10.1016/J.EGYPRO.2017.09.537.

[66] J. Maroušek, O. Strunecký, V. Bartoš, M. Vochozka, Revisiting competitiveness of hydrogen and algae biodiesel, Fuel 328 (2022) 125317. https://doi.org/10.1016/J.FUEL.2022.125317.
[67] B.K. Majhi, Cyanobacteria: Photosynthetic cell factories for biofuel production, J Bioresour Bioprod 10 (2025) 128–144. https://doi.org/10.1016/J.JOBAB.2024.10.001.
[68] Z. Zhou, D. Ouyang, D. Liu, X. Zhao, Oxidative pretreatment of lignocellulosic biomass for enzymatic hydrolysis: Progress and challenges, Bioresour Technol 367 (2023) 128208. https://doi.org/10.1016/J.BIORTECH.2022.128208.
[69] H.K. Sharma, C. Xu, W. Qin, Biological pretreatment of lignocellulosic biomass for biofuels and bioproducts: An overview, Waste Biomass Valor 10 (2019) 235–251. https://doi.org/10.1007/S12649-017-0059-Y/TABLES/3.
[70] Y. Yin, J. Wang, Production of biohydrogen, Biofuels and Biorefining: Volume 1: Current Technologies for Biomass Conversion (2022) 283–337. https://doi.org/10.1016/B978-0-12-824116-5.00002-7.
[71] J. Joy, C. Jose, P.L. Mathew, S. Thomas, M.N. Khalaf, Biological delignification of biomass, Green Polymers and Environmental Pollution Control (2016) 271–301. https://doi.org/10.1201/B19772-15/BIOLOGICAL-DELIGNIFICATION-BIOMASS-JITHIN-JOY-CINTIL-JOSE-LOVELY-MATHEW-SABU-THOMAS-MOAYAD-KHALAF.
[72] Z. Zhang, A.A. Donaldson, X. Ma, Advancements and future directions in enzyme technology for biomass conversion, Biotechnol Adv 30 (2012) 913–919. https://doi.org/10.1016/J.BIOTECHADV.2012.01.020.
[73] J.U. Hernández-Beltrán, I.O. Hernández-De Lira, M.M. Cruz-Santos, A. Saucedo-Luevanos, F. Hernández-Terán, N. Balagurusamy, Insight into pretreatment methods of lignocellulosic biomass to increase biogas yield: Current state, challenges, and opportunities, Appl Sci 9 (2019) 3721. https://doi.org/10.3390/APP9183721.
[74] D.M. Oliveira, T.R. Mota, A. Grandis, G.R. de Morais, R.C. de Lucas, M.L.T.M. Polizeli, R. Marchiosi, M.S. Buckeridge, O. Ferrarese-Filho, W.D. dos Santos, Lignin plays a key role in determining biomass recalcitrance in forage grasses, Renew Energy 147 (2020) 2206–2217. https://doi.org/10.1016/J.RENENE.2019.10.020.
[75] I. Cesarino, P. Araújo, A.P. Domingues, P. Mazzafera, An overview of lignin metabolism and its effect on biomass recalcitrance, Rev Bras Bot 35 (2012) 303–311. https://doi.org/10.1590/S0100-84042012000400003.
[76] S. Mussatto, Biomass Fractionation Technologies for a Lignocellulosic Feedstock Based Biorefinery (2016).
[77] A.K. Kumar, S. Sharma, Recent updates on different methods of pretreatment of lignocellulosic feedstocks: A review, Bioresou Bioprocess 4 (2017) 1–19. https://doi.org/10.1186/S40643-017-0137-9.
[78] K.O. Olatunji, D.M. Madyira, Comparative analysis of the effects of five pretreatment methods on morphological and methane yield of groundnut shells, Waste Biomass Valor 15 (2024) 469–486. https://doi.org/10.1007/S12649-023-02177-6/FIGURES/5.
[79] R.D. Prasad, A. Raturi, Prospects of sustainable biomass-based power generation in a small island country, J Clean Prod 318 (2021) 128519. https://doi.org/10.1016/J.JCLEPRO.2021.128519.
[80] E. Paone, A. Tursi, Production of biodiesel from biomass, Advances in Bioenergy and Microfluidic Applications (2021) 165–192. https://doi.org/10.1016/B978-0-12-821601-9.00006-6.
[81] K.O. Olatunji, D.M. Madyira, Enhancement of methane yield of Arachis hypogea shells using chemical pretreatment techniques, Mater Today Proc 105 (2024) 189–193. https://doi.org/10.1016/J.MATPR.2023.08.284.

[82] S. Anvari, R. Aguado, F. Jurado, M. Fendri, H. Zaier, A. Larbi, D. Vera, Analysis of agricultural waste/byproduct biomass potential for bioenergy: The case of Tunisia, Energy Sustain Dev 78 (2024) 101367. https://doi.org/10.1016/J.ESD.2023.101367.
[83] S.O. Dahunsi, S. Oranusi, V.E. Efeovbokhan, Pretreatment optimization, process control, mass and energy balances and economics of anaerobic co-digestion of Arachis hypogaea (Peanut) hull and poultry manure, Bioresour Technol 241 (2017) 454–464. https://doi.org/10.1016/J.BIORTECH.2017.05.152.
[84] K. Parmar, Biomass-An overview on composition characteristics and properties, IRA-Int J Appl Sci 7 (2017) 42–51. https://doi.org/10.21013/jas.v7.n1.p4.
[85] S. Sokhansanj, J. Fenton, Cost Benefit of Biomass Supply and Pre-Processing (2006).
[86] M.M. Tun, D. Juchelkova, M.M. Win, A.M. Thu, T. Puchor, Biomass energy: An overview of biomass sources, energy potential, and management in Southeast Asian Countries, Resources 8 (2019) 81. https://doi.org/10.3390/RESOURCES8020081.

2 Biorefinery of biomass and bioenergy

2.1 INTRODUCTION

Biomass is a carrier of renewable energy, known as bioenergy, and various bioenergy and chemical products can be produced, as illustrated in Figure 2.1 [1]. Bioenergy is a clean energy that can potentially substitute fossil fuels [2]. It is one of the renewable energy sources available that can help meet energy demand and support environmental remediation. This energy can generate heat, electricity, transportation fuels, and other products. Bioenergy is usually dispatchable and can be easily controlled, which allows the electricity grid operators to utilize energy from this source during off-peak demands [3]. Unlike other renewable energy sources, such as wind and solar, which depend on the weather, bioenergy is always accessible. For some seasonal biomass, bioenergy plants can still operate on other biomass sources for year-round power generation. Biomass can be converted to solid, liquid, or gas bioenergy, like biogas, biohydrogen, bioethanol, briquettes, biodiesel, etc. [2]. Biomass can be converted to bioenergy through direct conversion, biological, and thermochemical conversion technologies, as shown in Figure 2.2 [4]. The choice of conversion techniques mostly depends on the type and form of the available biomass [5]. Some of the bioenergy carriers produced from biomass are discussed below.

2.2 BIOGAS

Biogas production, achieved through a well-established technology known as anaerobic digestion, is regarded as one of the most sustainable energy sources [6]. Anaerobic digestion technology utilizes various types of organic biomass to produce biogas, and the resulting digestate is a nutrient-rich residue that can be employed as an organic fertilizer in agriculture [7,8]. Anaerobic digestion occurs in four stages: hydrolysis, acidogenesis, acetogenesis, and methanogenesis. Hydrolysis is the first stage of anaerobic digestion, where the biomass is broken down into smaller molecules for microorganism conversion, and it is regarded as the rate-limiting step. At this stage, carbohydrates are broken down into sugars, lipids into fatty acids, proteins into amino acids and peptides, and some portions into hydrogen and acetate. The products of the hydrolysis stage are converted into their respective organic acids, mainly by acidogenic bacteria during the acidogenesis stage. Additionally, hydrogen and carbon dioxide are released at this stage. Acetogenesis is the third stage, where acetogenic bacteria produce acetate from the carbon dioxide and hydrogen released from the previous stage. The last stage is the methanogenic stage, where biogas and methane are released from the products of acidogenesis and acetogenesis. The gas produced at this stage can be converted into electricity, heat, or hydrogen and stored as biomethane [7].

DOI: 10.1201/9781003633815-4

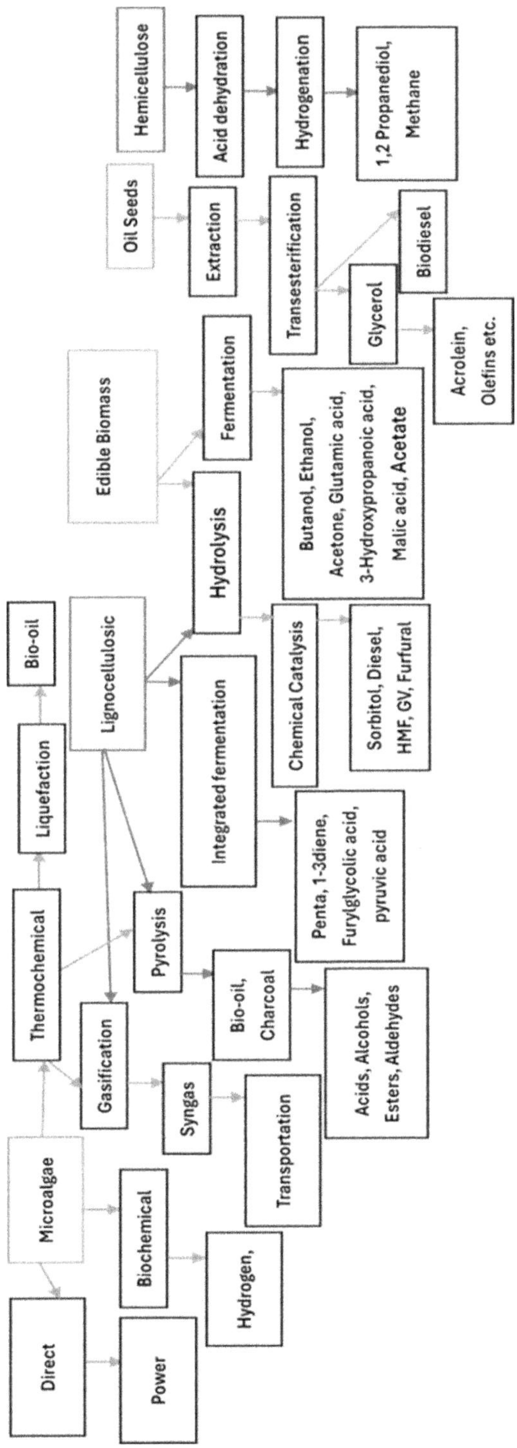

FIGURE 2.1 Biomass conversion routes to bioenergy and chemicals.

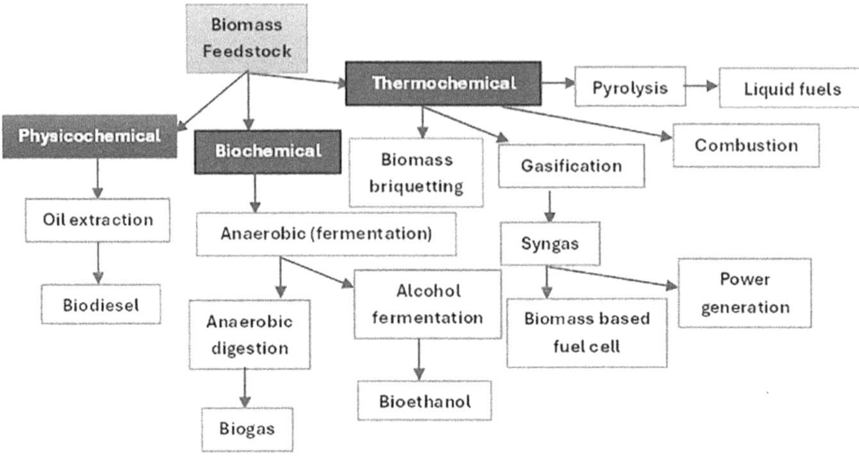

FIGURE 2.2 Biomass conversion technologies.

Biogas is a light hydrocarbon (LHC)-based renewable energy source that primarily consists of methane and carbon dioxide, along with other gases such as hydrogen sulfide, nitrogen, and siloxanes, in trace quantities. The composition of biogas depends on the constituents of the feedstock used and the conditions of the anaerobic digestion process [9]. Biogas and methane have lower heating values of 17 and 50 MJ/kg and higher heating values of 23.1 and 55 MJ/kg, respectively [7]. Biogas can be used for lighting, heating, cooking, and transportation fuel, and it can be purified and added to the grid. There are now more opportunities to benefit from sustainable energy due to the increasing interest in biogas generation, which also has numerous potential uses. As demand for biogas increases, regulations are being developed to regulate the process and optimize the use of currently available feedstock. A deeper understanding of the microbial community will play a significant role in enhancing biogas production technology. There are certainly intentions to transform the biogas plants into an innovative bioenergy industry that is safe and operates effectively [6]. Generally, biogas production is an eco-friendly means of generating renewable energy and a sustainable waste management technology that can help reduce the carbon footprint of the energy sector [7,10].

However, characteristics of biomass and production processes are the major challenges facing biogas production. Most of the biomass for biogas production is lignocellulose, and its production process is slow due to its low degradation rate. Lignocellulose consists of lignin, hemicellulose, and cellulose that are strongly intertwined, thereby restricting the activities of biogas-producing bacteria [11]. Therefore, pretreatment techniques must be applied before anaerobic digestion is required to enhance the biogas yield of these biomasses. Pretreatment techniques are classified into mechanical/physical, biological, chemical, thermal, nano-additives, and combined pretreatment techniques. These pretreatment techniques have been experimented with different biomasses, and it was observed that the effect is not universal [12,13]. Another challenge of biogas production from biomass is the significant differences in environmental effects among various production processes. The crucial stages that impact the ecological influence of biogas generation are biomass selection, production process, biogas application, and the management of the digestate. The biogas application and digestate management are determined mainly by local policies, and specific attention must be given to them during biogas production planning. Additionally, life-cycle analysis and social impact assessments should be developed to enhance the general acceptability and sustainability of biogas production. There is a need for a clearer understanding of environmental influences and social factors to facilitate a more holistic and cross-sectoral approach to site-specific life cycle assessment (LCA) [14].

2.3 BIOETHANOL

Bioethanol (CH_3CH_2OH) is a renewable biofuel identified as a potential alternative to fossil fuels in several countries [15]. This biofuel has a higher-octane number, faster flame speeds, and a higher heat of vaporization compared to gasoline, as well as wider flammability limits. These characteristics result in a leaner burn engine, higher compression ratio, and shorter burn time, which gives it higher theoretical efficiency merits than gasoline usage in the internal combustion engine [16]. It can be produced from various plant biomasses and can help reduce excessive greenhouse gas emissions. The quantity of carbon dioxide released during bioethanol burning is equivalent to the amount required for plant photosynthesis. Therefore, the CO_2 released during bioethanol production is considered biogenic [17]. Bioethanol can be applied by blending materials with gasoline to improve the octane and reduce CO_2 and other smog-causing emissions. Several biomass sources have been investigated as bioethanol feedstock and have shown good potential. Biomass with a considerable percentage of sugar content can serve as feedstock for bioethanol production. The most acceptable feedstocks for bioethanol generation are categorized as sucrose-based (e.g., sweet sorghum, sugarcane, sugar beet) and starchy (e.g., rye, cassava, corn, wheat) [18,19]. The use of lignocellulose biomass as a bioethanol feedstock has been investigated for the medium and long term due to its abundance and low cost [20].

Bioethanol production from lignocellulose sources is an attractive and sustainable technique, as it utilizes renewable resources and does not compete with the food supply [21]. It can be sourced through different agricultural and forestry residues or directly harvested from the forest [22]. Lignocellulose biomass for bioethanol production is grouped into six major groups, which are hard wood (polar, aspen), crop residues (rice hulls, cane, corn stover, pulp, sweet sorghum bagasse, etc.), softwood (spruce, pine), herbaceous biomass (switchgrass, alfalfa hay, etc.), cellulose wastes (recycled paper sludge, waste paper,), and municipal solid wastes [23]. Generally, the constituents of lignocellulose biomass are 27% lignin, 43% cellulose, 20% hemicellulose, and other components are 10% [22]. Variations in the composition of lignocellulose biomass can be a merit to the process (broader range of feedstocks and more product availability) and, at the same time, a limitation (need for more technologies) [24]. Biomass with heterogeneous morphological arrangements needs more complex chemical processes compared to uniform and consistent biomass during conversion technology [25]. Bioethanol is usually produced through the fermentation of sugar or starch-based feedstocks. The technology also depends on the type of starch- or sugar-based feedstock, but some stages remain the same regardless of the feedstock type [26]. Five major steps are required in bioethanol production: biomass generation, conversion to usable fermentation feedstock, microbial fermentation to release bioethanol, utilization of unfermented residue, and environmental management of the process [18].

During bioethanol production, feedstock received at the processing plant must be carefully stored in the right conditions to avoid bacterial contamination and early fermentation [27]. During bioethanol production from lignocellulose biomass, the stages mostly used are the following: (i) biomass pretreatment for easy accessibility, (ii) enzymatic or acid hydrolysis of polysaccharides into simple sugars, (iii) microbial fermentation of simple sugars (pentoses and hexoses) to ethanol, and (iv) bioethanol recovery, as illustrated in Figure 2.3 [28]. Pretreatment application facilitates the conversion of complex sugars into simple sugars, depending on the biomass composition and treatment technique employed. Saccharomyces cerevisiae is the most used microorganism during bioethanol production from sugar-based feedstocks because of its strength in degrading sucrose to hexose (fructose and glucose). Aeration is an essential process input when using *S. cerevisiae*, as its cells require small amounts of oxygen to synthesize fatty acids and sterols during bioethanol production. This microorganism has a low tolerance to higher salt and sugar concentrations in the medium, as well as higher temperatures [29]. However, when experimented in bioethanol production, the yeast Schizosaccharomyces pombe has a high tolerance to high salt concentrations and solid content [29]. The use of other microorganisms, such as *Mucor indicus, Zymomonas mobilis, Thermoanaerobacter ethanolicus, Klebsiella oxytoca, Candida shehatae, Escherichia coli, and Pichia stipites,* has been investigated; however, it is not as effective as *S. cerevisiae* [30,31].

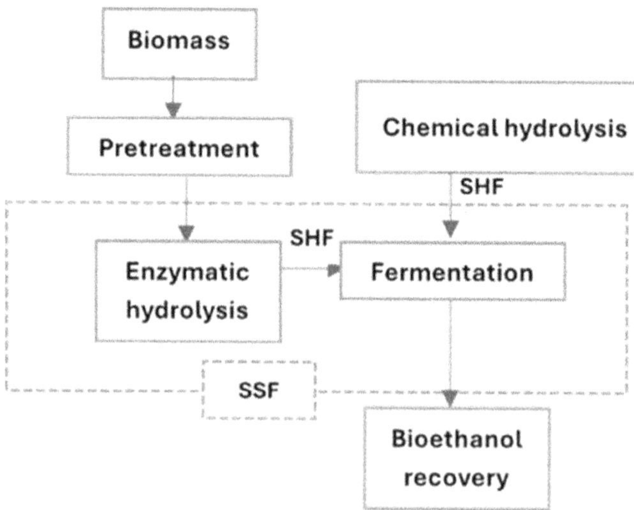

FIGURE 2.3 Schematic diagram of ethanol production from biomass in a general principle.

Bioethanol as bioenergy is mostly used in transportation when mixed with gasoline or as an octane improver (ethyl tertiary butyl ether (ETBE), a mixture of 45% bioethanol and 555 isobutylene). Some countries use ETBE as an alternative to methyl tertiary butyl ether (MTBE) as an octane improver. Still, it is prohibited in countries like Canada and the USA because the emissions are carcinogenic [18]. E10, a mixture of 10% ethanol and 90% gasoline, is the most widely accepted bioethanol mix, and up to 15% ethanol and 85% gasoline (E15) is permitted for conventional gasoline-powered vehicles in the USA. E15 can be used in flexible fuel vehicles (FFVs) depending on the season and geography, but E10 can be used without any engine modifications. Approximately 97% of gasoline available in the USA contains some percentage of bioethanol [32]. Some of the limitations to the use of bioethanol include its corrosion properties (hygroscopic nature), which can damage the injector and electric fuel pump, engine ignition challenges during cold weather, emulsion formation, and tribological impacts on lubricant characteristics and engine oil performance [18]. An engine's bioethanol performance can be improved by reducing the mass of engine parts, using laser texture, optimizing lubricant composition, and applying coatings, as well as by increasing its lifespan by reducing friction and wear. Applying synthetic oil is another way to solve these challenges [33].

2.4 BIOHYDROGEN

Hydrogen is a zero- or low-carbon energy source considered a bright and clean energy source of the 21st century. It is a major raw material for several chemical processes and a source of clean energy for different power production technologies. With the growing interest in hydrogen production and applications, global demand for hydrogen is expected to exceed 300 million metric tons by 2050 [34]. Hydrogen energy will play a significant role in economic and environmental development. It is a clean fuel whose combustion produces only water vapor as a by-product [35]. There are various benefits of hydrogen as a fuel, including zero emissions, versatility, renewability, widespread distribution, rapid recovery, high conversion efficiency, and cleanliness, which make it an excellent energy source among others [36–38]. Due to this, it is considered the most environmentally benign and bright energy that can replace fossil fuels. Hydrogen has been mostly used in the transport and fuel sector for several years. Biomass and coal are the primary raw materials for hydrogen production, and they have become a major focus of interest in industrial hydrogen energy development. Presently, a large quantity of hydrogen is produced globally through steam methane conversion from fossil fuels. Other technologies that rely on renewable sources, such as water splitting and hydrogen production from biomass, particularly from lignocellulose substrates, are currently at the pilot-scale stage or the industrial level [39]. Hydrogen generation from biomass can increase its availability, support the economy, enhance flexibility, and reduce greenhouse gas emissions [40]. The amount of carbon dioxide released during the combustion of hydrogen produced from biomass is equivalent to the carbon consumed by the plant during photosynthesis. Therefore, hydrogen from biomass is regarded as carbon-neutral energy [41]. Different types of technology are being used to produce hydrogen from biomass. These technologies are classified as biological conversions,

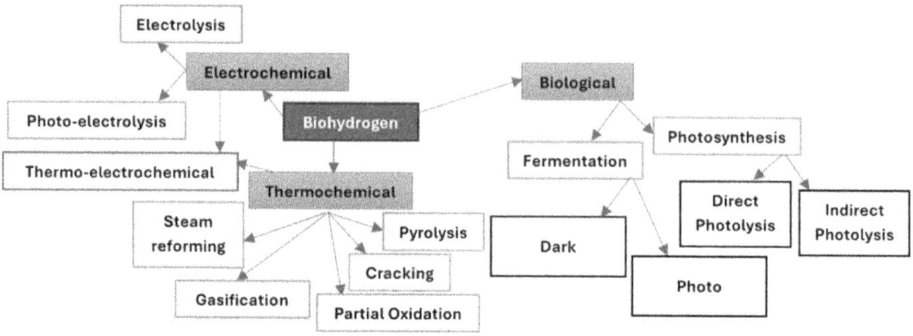

FIGURE 2.4 Biohydrogen production technologies.

thermochemical technologies, and potentially electrochemically assisted generation, as illustrated in Figure 2.4 [42].

Thermochemical conversion of biomass into hydrogen is considered one of the most advanced technologies. It originated using similar techniques to those employed for biofuels, such as biomethane, which utilizes steam methane reforming (SMR) technology [43]. Gasification, pyrolysis, and aqueous phase reforming (APR) are the three major thermochemical technologies for biohydrogen production. The gasification technique is a highly endothermic process that occurs under limited oxygen at a temperature of approximately 1000 °C. This method depletes the oxidizing agent to release syngas, a synthetic gas composed of hydrogen, nitrogen, methane, carbon dioxide, and carbon monoxide. The techniques depend on the oxidizing agent utilized and can be regarded as steam, air, or oxygen gasification [44]. Additionally, traces of both inorganic and organic pollutants were detected in the syngas. These pollutants include LHCs, such as tar, methane, and inorganic impurities like ammonia, hydrogen sulfide, hydrochloric acid, and alkali metals [45]. During the process, the moist biomass is placed in the gasifier to dry before undergoing a pyrolysis reaction, which releases gaseous compounds (CO_2, CO, LHC) and tar. After this stage, a reduction reaction occurs, where the combustion of the char (solid residue from the combusted biomass) releases hydrogen. The process can be represented as shown in Equation 2.1 [46]. The residue from the process is moved to the combustor to produce carbon dioxide, which can generate heat for the system.

$$\begin{aligned} Biomass + Air \rightarrow H_2 + CH_4 + CO + CO_2 + N_2 + CH_4 + \\ H_2O + tar + LHC + char. \end{aligned} \tag{2.1}$$

Another thermochemical conversion technique for biohydrogen from biomass is pyrolysis. This technique is similar to gasification but can be performed without an oxidizing agent and at lower temperatures [47]. Although a small quantity of oxidizing agents can be used to facilitate heat production. The technique can take place at temperatures between 400 and 800 °C and a pressure of around 5 bar. Based on the temperature used, pyrolysis is divided into flash, slow (conventional), and fast pyrolysis. The process of biohydrogen production through pyrolysis is illustrated in Equation 2.2 [48].

$$biomass + heat \rightarrow H_2 + CH_4 + CO_2 + CO + H_2O + Bio - oil + charcoal\ (\Delta H^o > 0\, kJ\,/\,mol). \tag{2.2}$$

APR is another thermochemical technique of hydrogen production from biomass, as illustrated in Figure 2.5. It converts primarily oxygenated materials into hydrogen. Biomass molecules are solubilized during the aqueous stage and react with water molecules at high pressures (about 50 bar) and low temperatures (< 270 °C) [49]. The technique is most efficient with biomass-derived oxygenated hydrocarbons that have a C:O ratio of 1:1 and are soluble in water, like glycerol, methanol, glucose, ethylene glycol, or polyols like sorbitol [39]. The technology is more efficient when the biomass is pretreated before the APR process.

The biological conversion technique is another novel method for producing hydrogen from biomass, and it is expected to revolutionize the biorefinery system. The technology is reported to be infinite and environmentally friendly because it is a carbon-neutral technology [50]. Biological hydrogen production techniques can be divided into fermentation, the enzymatic process, biophotolysis, and microbial electrolysis. The fermentation technique is divided into two types, depending on the need for light during conversion: dark fermentation (no light required) and photofermentation (requiring light). Photosynthetic bacteria used solar energy to release hydrogen from biomass during the photofermentation process [49]. Biophotolysis can be divided into two categories: indirect and direct biophotolysis. Each of these techniques depends on the type of enzyme used for hydrogen release catalysis [39]. Generally, microorganisms such as cyanobacteria and microalgae produce the enzymes necessary for hydrogen synthesis, including hydrogenases and nitrogenases [51]. Hydrogenase enzymes reversibly catalyze the transformation of protons to hydrogen, while nitrogenases reduce protons (H^+) from adenosine triphosphate

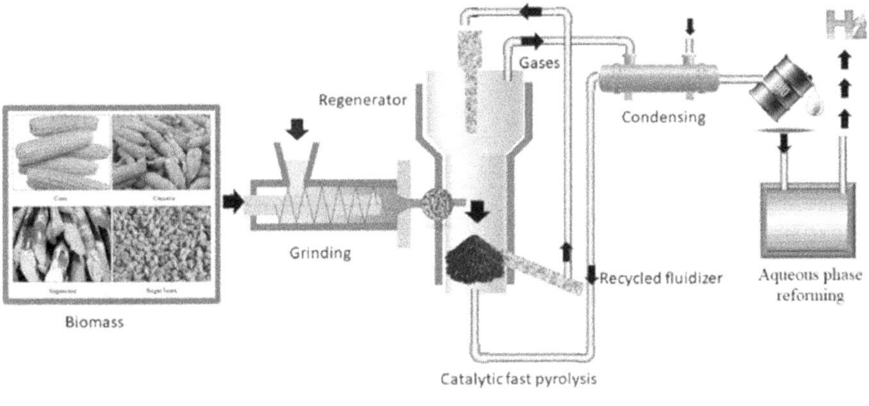

FIGURE 2.5 Schematic illustration of aqueous phase reforming for biohydrogen production [49].

and electrons to produce hydrogen. This technique is carried out at lower pressures (1 atm) and temperatures (30–60 °C), reducing the energy cost compared to the thermochemical conversion process. Furthermore, the microorganisms utilized can be easily recovered through replication, thereby reducing the cost compared to chemical catalysts that are degraded in thermochemical processes. A biological technique primarily interested in waste management encourages the conversion of all forms of biomass to biohydrogen.

Hydrogen can also be generated from biomass through an electrochemical technique, as presented in Figure 2.6. The primary difference between biomass and water electrolysis lies in the reaction that occurs at the anode; instead of producing oxygen from water, biomass is oxidized [49]. Biomass can be electrolyzed to hydrogen through microbial electrolysis cells (MECs) and proton exchange membrane electrolysis cells (PEMECs). Both methods are routinely used for biobased materials, such as glycerol and ethanol. It is not easy for electrolysis to directly convert polymeric compounds like wood sawdust or cellulose; therefore, a single or double chamber with cathode/anode coupling is needed. The organic content of the biomass is converted through an oxidation process at the anode to produce protons (H^+), and a reduction takes place at the cathode to allow hydrogen production [52]. MEC is the latest cleaner biochemical method to generate biohydrogen from biomass. MECs are composed of an anode and a cathode electrode in anaerobic conditions and are connected to a power source [53]. The technology is similar to microbial fuel cells (MFCs); their electrodes and reaction setup are similar. The major difference between the two is that air is not required for the MEC's cathode, and the outputs are not the same [54]. The MFCs generate electricity directly, whereas MECs release biohydrogen on the outer part of the cathode. For the MEC technique,

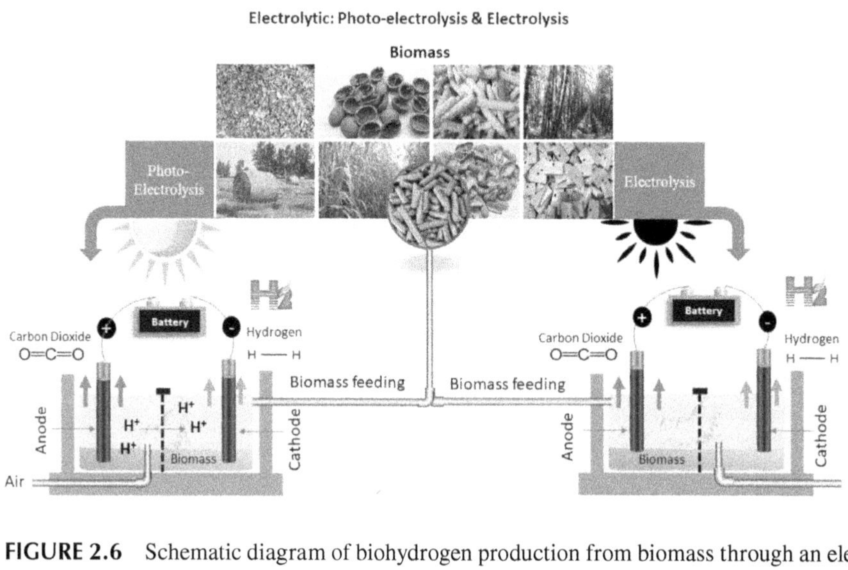

FIGURE 2.6 Schematic diagram of biohydrogen production from biomass through an electrochemical process [49].

biomass oxidation to release biohydrogen is performed with electrochemically active microbes as catalysts, which makes the technique biochemical. The PEMEC technique has a polymer membrane that converts protons to release a high percentage of hydrogen gas [55]. Different techniques for biohydrogen production have been investigated, and it has been established that the major limitations to the technology are conversion efficiency, biomass type, and the requirement to safely integrate generation with purification and storage technologies.

2.5 BIODIESEL

Biodiesel is a renewable biofuel composed of fatty acid alkyl esters produced through transesterification processes, which can serve as a substitute for traditional diesel. It is a fuel that can contribute to global energy security due to its multiple advantages, which include low greenhouse gas emissions, satisfactory performance in combustion engines, biodegradability, sustainability, and non-toxicity [56]. The production process is chemical and differs from that of bioethanol. It can be produced from biomass (seeds and oil plants), and the process depends on oil extraction and conversion to biodiesel by breaking down the long-chain fatty acids to glycerol and substituting it with methanol during transesterification [57]. The process is divided into four conventional techniques: dilution or direct blending, microemulsion, pyrolysis/thermal cracking, and transesterification [58]. Biodiesel can be generated from various biomass sources, including jatropha seed, sandbox seed, breadfruit, ackee, yellow oleander, Kariya seed, palm kernel seed, and algae [59–61]. Production from lignocellulose involves four critical stages: delignification, saccharification, fermentation with oleaginous microorganisms (Oms) for enhanced lipid synthesis, and subsequent conversion through esterification [60]. Biomass can be used in two different paths during biodiesel production. It can produce lipids through the hydrolysis of hemicellulose and cellulose, which can be further processed to generate biodiesel. Another pathway involves producing economical and eco-friendly catalysis from biomass, such as stalks, stems, and leaves, to improve the conversion rate during transesterification [62]. Transesterification can also be catalyzed by hydroxides and alkoxides of metals, yielding glycerol and alkyl esters as primary products, with biodiesel as the remaining product. Recently, various methods have been employed for biodiesel production through transesterification reactions [63]. One of the major benefits of biodiesel is that it can be used in an existing engine without modification. Biodiesel can be mixed with petroleum diesel up to 20% and still achieve optimal engine performance [64].

Biodiesel is a bioenergy fuel used for heating and automotive purposes, either pure or mixed with petrol-diesel. The feedstock can be grown on marginal or abandoned lands and does not compete with the food supply. The major benefit of biodiesel compared to other traditional petrochemical fuels is its carbon neutrality and ability to reduce CO_2 release. It is nearly free of sulfur ($SO_2 < 0.001\%$) with a higher flash point, and its application promotes lower reductions in fine dust, greenhouse gas emissions, and unburnt particulates in the atmosphere [62]. A blend of biodiesel provides superior lubricating characteristics; fuel lubricity can be improved by approximately 65% when 1% of biodiesel is blended with petroleum diesel fuel. This can reduce engine wear and extend the lifespan of the fuel injector. Nearly all diesel engines on and off-road vehicles, as well as marine vessels, can run on a blend of biodiesel. Its use is primarily advisable in marine environments due to its ability to eliminate potential contamination of waterways. Biodiesel can also be used for power generation on stationary engines, and it is an attractive option to meet power regulations in terms of emissions control that can be achieved through petroleum diesel. With little modification to the burner train, biodiesel can replace expensive natural gas as boiler fuel. Due to its non-toxicity and biodegradability, it can be utilized as a carrier for fertilizers and pesticides in agriculture. The application of biodiesel as an industrial solvent to replace high-volatile organic compound (VOC)-containing petroleum solvents has been proven to be an attractive option in industry and as a household cleaning agent. Biodiesel can be used as a fuel additive to keep the pumps, injectors, and other combustion parts clean [65]. The microstructural arrangement of biomass is a significant limitation to biodiesel production from these feedstocks. It is, therefore, essential that biomass is pretreated through various techniques before biodiesel production, which increases the production cost [25]. Another major challenge of biodiesel application is the viscosity, which can be improved by introducing some additives. It has a solvent effect that causes deposit accumulation on the tanks and pipes of previously used petrol-diesel equipment. These deposits can clog the initial filters; therefore, precautions are needed. With time, biodiesel softens and degrades specific types of natural rubber and elastomer compounds. Additionally, materials such as lead, bronze, zinc, brass, and copper may oxidize and generate sediments. It is advisable to replace all components incompatible with biodiesel with either steel or aluminum when introducing biodiesel to the system [65].

2.6 BRIQUETTE

Briquettes are bioenergy biofuels produced from flammable biomass through the briquetting process. It is a renewable energy source derived from biomass and can help reduce greenhouse gas emissions while generating heat and electricity [66]. Briquetting is the process of densifying biomass into fuel logs using a briquette machine, primarily made from agricultural and forest residues. Biomass densification into fuel briquettes is one method used to convert biomass into bioenergy. This technique provides uniform solid fuel, improved physical and energy properties, facilitates transport and storage, and produces more homogeneous combustion [67]. It enhances these properties, resulting in solid fuel that can be stored, handled, and transported. The biomass densification technique is an effective method that guarantees desirable properties for solid biofuels, such as high energetic density, resistance, higher apparent density, and hardness. The process can be carried out in the form of briquettes, cubes, pellets, or disks [68]. Energy conversion in briquettes occurs through direct burning and can be utilized for cooking and heating purposes at both domestic and industrial scales [69]. Briquettes, a form of fuel derived from biomass, have been adopted in both developed and developing countries [70]. Fuel briquettes are classified into two categories: carbonized and non-carbonized briquettes. The decision to carbonize or not depends on the availability of biomass, the technology available, and the existing or anticipated market environment [66].

As illustrated in Figure 2.7, the briquette production process mainly includes biomass acquisition, processing, and densification. During densification, pressure, heat, and binding agents are applied to biomass to produce briquettes. This compacts the loose biomass to reduce its volume and agglomerates the feedstock, resulting in a final product in a compressed state [71]. Biomass densification into briquettes typically begins with sorting and cleaning, which removes impurities from the feedstock. Feedstock drying and size reduction may be required depending on the moisture content and size of the biomass. These can be done using different drying and size reduction techniques, but they must be economically viable. Drying the feedstock when its moisture content is high is essential to increase efficiency, but it should not be dried excessively. However, some feedstocks, such as groundnut shells, rice husks, and coffee husks, may not require drying. Binders can be used during the feedstock mixture or after carbonization before densification. Some biomasses cannot aggregate unless a binder is added, particularly when the low-pressure compaction method is used [72]. Binders can be organic (e.g., molasses, cassava paste, cow dung, wastepaper pulp, and starch), inorganic (e.g., sodium silicate, clay, cement, lime, and plaster), or compound (e.g., polymer binder, tar, lignosulfonate binder, pitch, and petroleum bitumen binder) [71]. Briquettes can vary in quality and size, depending on the feedstock, mold, and technologies used during production. It can be square, cylindrical, rectangular, or polygonal in shape. The qualities of a good briquette include good combustion, durability during storage and handling, stability, and environmental safety when combusted [73]. Calorific value, ease of ignition, and percentage ash are the parameters that determine the combustion properties of a briquette. At the same time, the environmental impact depends on the toxic emissions produced during the combustion process. Briquettes are an eco-friendly, renewable energy source that can assist in energy security and environmental management.

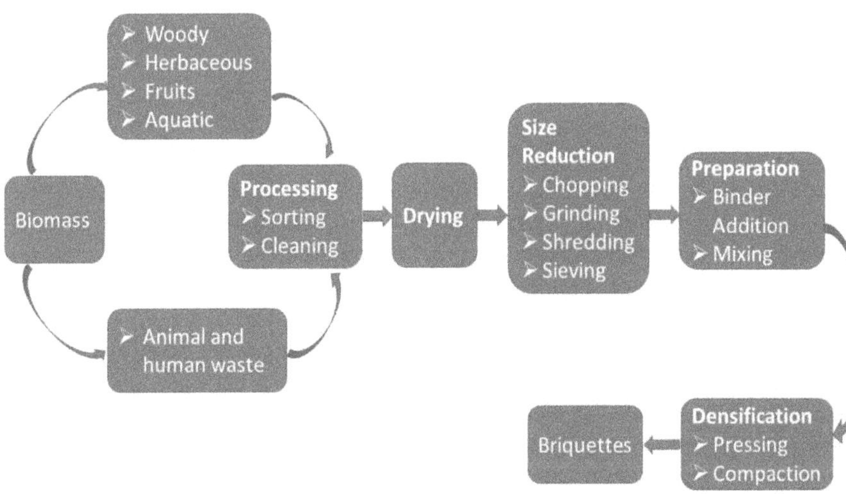

FIGURE 2.7 Briquette production processes [71].

However, there are some limitations to the application of its full potential. Briquetting is relatively new in some developing countries, and a lack of technological capacity hinders the production of densification equipment. Adopting biomass densification depends on biomass availability, available technologies, and the briquette market [74]. Feedstock availability is not a challenge in developing countries, but optimizing the mechanical and chemical treatments required for most available biomass is a challenge. Some technologies that produce better quality require high investment costs and high energy input. To encourage the production of briquettes in large quantities, especially in developing countries, adequate financing, research, and development of more user-friendly and economical technologies are necessary.

2.7 PYROLYSIS/BIO-OIL

Pyrolysis is the thermochemical conversion of biomass into bioenergy, a green energy that can substitute fossil fuels. It involves breaking down biodegradable biomass at high temperatures without full combustion. The process decomposes the polymer chain in biomass macromolecules to produce high-energy-density fuel [75]. The end products of the pyrolysis process are biochar, pyrolysis oil, and non-condensable gases. The quality and quantity of the process depend largely on biomass, heating rate, particle size, gas residence time, temperature, reactor, pretreatment techniques, catalysis, and atmosphere [76]. It has been observed that pyrolysis oil generated from biomass contains impurities and cannot be used directly as a replacement for fossil fuels. Therefore, a different catalytic process is required to purify the bio-oil and produce pure pyrolysis oil, as well as other value-added chemicals [77]. Pyrolysis oil from biomass is a constituent of aromatic, naphthenic hydrocarbons, aliphatic, and oxygenated compounds like furans, alcohols, ethers, aldehydes, phenols, acids, and ketones. It has better energy density than biomass and can be stored, transported, and used as a renewable liquid biofuel or chemical generation [78]. Although pyrolysis oil has merits of processing and transportation, its complex constituents, poor stability, high viscosity, and corrosion properties limit its direct application potential [79].

Biomass pyrolysis can be classified into three major types: slow, fast, and flash pyrolysis, depending on the operational parameters used. A novel type of pyrolysis, investigated as an intermediate between slow and fast pyrolysis, has been developed. The four types of pyrolysis are distinguished by their heating rates, residence times, and temperatures [80]. Advanced pyrolysis techniques are mostly modified traditional processes that enhance the process's characteristics and outputs. Several techniques of the pyrolysis process exist, including co-pyrolysis, hydropyrolysis, catalytic pyrolysis, and microwave pyrolysis [79]. The chemical structure of biomass is altered during heating under anoxic or anaerobic conditions, releasing volatile compounds and realigning the residues. When the heating continues, the unstable, volatile compounds generated in the initial reaction will be subsequently converted into a rich and different pyrolysis product. Therefore, biomass pyrolysis conversion technology can be categorized into primary and secondary conversion [76]. Pyrolysis oil has been utilized in combustion boilers, turbines, as a fuel for engines, as a transportation fuel, and as a renewable substrate for other chemicals [81]. However, biomass pyrolysis has limitations, including complex composition, poor conversion efficiency, and difficulty in separating and purifying, which hinder the commercial production of pyrolysis oil [82]. Studies have shown that adding catalysts during pyrolysis can alter the distribution and characteristics of the pyrolysis oil, enhance process efficiency, and make the process more economical [83].

2.8 SYNGAS

Syngas (synthetic natural gas) is a gaseous substance that can be produced during the pyrolysis or gasification of biomass. Gasification is a thermochemical process that converts biomass into multifunctional gaseous constituents, known as syngas, and other value-added products [84]. It is a mixture of different constituents, including CO_2, CO, CH_4, and H_2, with some impurities such as NH_3, SO_2, NO_2, H_2S, tar, and low-carbon alcohols as presented in Equation 2.3 [79]. Additionally, a solid residue, referred to as char, is found mostly after gasification. The equation illustrates the biomass as a simplified gasification reaction process [85]. The composition of syngas depends on the feedstock used, as well as other process parameters, such as gasifier temperature and pressure, gasification technology, and type of bed materials [86]. Gasification can be divided into four stages: drying (endothermic), pyrolysis (endothermic), oxidation (exothermic), and reduction (endothermic). A tar purification stage can be added to generate LHCs from the longer tar molecules [85]. Syngas produced from biomass can be utilized for electricity or heat generation, as well as for the production of chemicals (such as ammonia) and as a raw material for the thermochemical production of second-generation biofuels, including ethanol, methanol, Fischer-Tropsch diesel, and dimethyl ether (DME) [87]. Syngas production as an energy carrier is economical because the same equipment, storage, and transportation used for liquefied petroleum gas can be used. Still, technology is of more concern, especially in developing countries.

$$\begin{aligned} Biomass \rightarrow CO + H_2 + CO_2 + CH_4 + H_2O + H_2S + \\ NH_3 + C_xH_y + Tar + Char. \end{aligned} \tag{2.3}$$

2.9 BIOCHAR

Biochar is a rich-carbon and dense material produced through the thermochemical degradation of biomass and other carbonaceous materials under the control or absence of oxygen. It is formed when intermolecular and intramolecular arrangements of biomass occur, and the volatile water is released. It can be produced with different technologies such as torrefaction, pyrolysis, hydrothermal carbonization, and gasification [88]. Biochar production primarily aims to release energy or reduce the volume of waste materials. Biochar applications are gaining more attention in different sectors like energy, environment, and agriculture because of their simplicity, availability of raw materials, and improved physicochemical characteristics [89]. It is a material that can potentially remove contaminants from wastewater and can improve soil fertility, water retention, and nutrient absorption in agriculture. It is an economic product that transforms waste into wealth. When suitable feedstock is processed at optimum process parameters, it produces biochar with excellent chemical and physical properties, which include good porosity, rich functional groups, stable structural characteristics of minerals and organic carbon, which makes it acceptable as a renewable energy fuel, soil enhancer, pollution adsorption, and other environmental management and ecological protection [90].

The selection of bioenergy technologies depends on several factors, and the use of LCA will play a crucial role in evaluating these factors. This tool can assist in examining the process contribution to the environment, net energy, water requirements, and other crucial information [91]. LCA analysis of ethanol production from corn, considering water requirements, greenhouse gas release, and net energy, shows a significant reduction of about 40–50% in emissions compared to gasoline [92]. The use of LCA to examine the environmental impact of ethanol production from cassava pulp and its application as a transportation fuel was considered. It was observed that the most sustainable scenario involves utilizing biogas fuel as an alternative to fuel oil during ethanol production, along with allocating cassava pulp according to its economic value. Analysis of its application as fuel for vehicles shows a reduction in greenhouse gas emissions with 6, 10, and 48% reduction for 10, 20, and 85% of ethanol mixed with gasoline, respectively [93]. The study also noted a significant reduction in the fossil fuel depletion rate of 14, 19, and 59% for the same blending. The lower mixture of ethanol and gasoline has a slightly varying influence on human toxicity (HT), terrestrial acidification (TA), and photochemical oxidation formation (POF) compared to gasoline. However, a mixture with higher ethanol resulted in a stronger negative influence on POF, HT, TA, and freshwater eutrophication [93].

2.10 BIOMASS BIOREFINERY INDUSTRIES

There are examples of lignocellulose biomass biorefineries that are in operation globally, with others under construction. Table 2.1 presents some industries that utilize hemicellulose and cellulose primarily for biofuel production, and lignin is used for energy production [94]. It is expected that this scenario will change in the future due to the recent incentives assigned to biomass valorization for producing valuable products, such as phenolic (aromatic) compounds.

2.11 CONCLUSION

Biomass presents a promising source of feedstock for biorefineries due to its renewability, versatility, and minimal environmental impact. Green biorefinery of biomass offers a solution to produce various energy and value-added products without direct competition with the food supply chain by utilizing waste and residues. Specific attention has been given to lignocellulose biomass, including agricultural residue, energy crops, food waste, municipal solid waste, and animal manure, for the production of bioenergy. Biorefinery of these second-generation feedstocks is gaining attention to enhance living standards and support the circular economy. Progress has been made in the biorefinery of biomass to produce bioenergy that can substitute

TABLE 2.1
Examples of biomass biorefinery industries

Industry	Scale	Feedstock	Fractionalization technique	Primary product	Lignin utilization
Poet-DSM (United States)	Commercial	Agricultural residues	Enzymatic hydrolysis	Bioethanol and biogas	Fuel
Beta renewables (Italy)	Commercial	Agricultural residues	Steam explosion	Bioethanol	Fuel
Abengoa Bioenergia (United States)	Commercial	Switchgrass, corn stover, and wheat straw	Steam explosion	Bioethanol	Fuel for electricity and steam
Chempolis (Finland)	Demonstration	Agricultural residues	Organosolv	Bioethanol and cellulose	Fuel
Abengoa Bioenergia (Spain)	Demonstration	Wheat straw	Steam explosion	Bioethanol	Fuel and feed additive
Inbicon (Denmark)	Demonstration	Agricultural residues	Hydrothermal	Bioethanol	Fuel
CIMV (France)	Pilot	Agricultural residues (rice and wheat straw, hardwoods)	Organic acid and organosolv	Cellulose, C5, lignin, and sugars	Performance materials
Clariant (Germany)	Pilot	Wheat straw	Pressurized steam treatment and enzymatic hydrolysis	Bioethanol	Fuel
Biorefinery demo plant (Sweden)	Pilot	Wood chips. Wheat straw, bagasse, corn stover, and energy grass	Dilute acid and enzymatic hydrolysis	Bioethanol	Fuel and performance materials

Adapted from [95].

for fossil fuels and assist in sustainable waste management. However, several limitations are still delaying its industrial application. From the economic point of view, optimization of the technologies for bioenergy generation needs to be optimized for efficient production. Most of the biomass is lignocellulose, which necessitates the adoption of proper pretreatment techniques to improve the economic viability of biorefineries.

REFERENCES

[1] S.K. Bardhan, S. Gupta, M.E. Gorman, M.A. Haider, Biorenewable chemicals: Feedstocks, technologies and the conflict with food production, Renew Sustain Energy Rev 51 (2015) 506–520. https://doi.org/10.1016/J.RSER.2015.06.013.
[2] L. Zhao, Z.F. Sun, C.C. Zhang, J. Nan, N.Q. Ren, D.J. Lee, C. Chen, Advances in pretreatment of lignocellulosic biomass for bioenergy production: Challenges and perspectives, Bioresour Technol 343 (2022) 126123. https://doi.org/10.1016/J.BIORTECH.2021.126123.
[3] M. Bijarchiyan, H. Sahebi, S. Mirzamohammadi, A sustainable biomass network design model for bioenergy production by anaerobic digestion technology: Using agricultural residues and livestock manure, Energy Sustain Soc 10 (2020) 1–17. https://doi.org/10.1186/S13705-020-00252-7/FIGURES/6.
[4] N.B. Singh, A. Kumar, S. Rai, Potential production of bioenergy from biomass in an Indian perspective, Ren Sustain Energy Rev 39 (2014) 65–78. https://doi.org/10.1016/J.RSER.2014.07.110.
[5] M.A. Mehmood, M. Ibrahim, U. Rashid, M. Nawaz, S. Ali, A. Hussain, M. Gull, Biomass production for bioenergy using marginal lands, Sustain Prod Consum 9 (2017) 3–21. https://doi.org/10.1016/J.SPC.2016.08.003.
[6] M. Arshad, S. Javed, A.R. Ansari, A. Fatima, M.I. Shahzad, Biogas: A promising clean energy technology, Bioenergy Resources and Technologies (2021) 91–120. https://doi.org/10.1016/B978-0-12-822525-7.00011-1.
[7] P. Gupta, C. Kurien, M. Mittal, Biogas (a promising bioenergy source): A critical review on the potential of biogas as a sustainable energy source for gaseous fuelled spark ignition engines, Int J Hydrogen Energy 48 (2023) 7747–7769. https://doi.org/10.1016/J.IJHYDENE.2022.11.195.
[8] D.M. Madyira, K.O. Olatunji, Enhancing methane yield from duck waste by co-digestion with Xyris capensis, E3S Web Conf 433 (2023) 01002. https://doi.org/10.1051/E3SCONF/202343301002.
[9] R. Rajkumar, C. Kurinjimalar, Food wastes/residues: Valuable source of energy in circular economy, Handbook of Biofuels (2022) 147–163. https://doi.org/10.1016/B978-0-12-822810-4.00007-5.
[10] K.O. Olatunji, D.M. Madyira, Enhancing the biomethane yield of groundnut shells using deep eutectic solvents for sustainable energy production, Front Energy Res 12 (2024) 1346764. https://doi.org/10.3389/FENRG.2024.1346764.
[11] K.O. Olatunji, D.M. Madyira, Comparative analysis of the effects of five pretreatment methods on morphological and methane yield of groundnut shells, Waste Biomass Valor 1 (2023) 1–18. https://doi.org/10.1007/S12649-023-02177-6/FIGURES/5.
[12] K.O. Olatunji, N.A. Ahmed, O. Ogunkunle, Optimization of biogas yield from lignocellulosic materials with different pretreatment methods: A review, Biotechnol Biofuels 14 (2021) 1–34. https://doi.org/10.1186/S13068-021-02012-X.
[13] H. Rabemanolontsoa, S. Saka, Various pretreatments of lignocellulosics, Bioresour Technol 199 (2016) 83–91. https://doi.org/10.1016/j.biortech.2015.08.029.

[14] S. Huttunen, K. Manninen, P. Leskinen, Combining biogas LCA reviews with stakeholder interviews to analyse life cycle impacts at a practical level, J Clean Prod 80 (2014) 5–16. https://doi.org/10.1016/J.JCLEPRO.2014.05.081.
[15] C. Manochio, B.R. Andrade, R.P. Rodriguez, B.S. Moraes, Ethanol from biomass: A comparative overview, Renew Sustain Energy Rev 80 (2017) 743–755. https://doi.org/10.1016/J.RSER.2017.05.063.
[16] S. Barua, D. Sahu, F. Sultana, S. Baruah, S. Mahapatra, Bioethanol, internal combustion engines and the development of zero-waste biorefineries: An approach towards sustainable motor spirit, RSC Sustain 1 (2023) 1065–1084. https://doi.org/10.1039/D3SU00080J.
[17] S. Bastianoni, N. Marchettini, Ethanol production from biomass: Analysis of process efficiency and sustainability, Biomass Bioenergy 11 (1996) 411–418. https://doi.org/10.1016/S0961-9534(96)00037-2.
[18] A. Bušić, N. Mardetko, S. Kundas, G. Morzak, H. Belskaya, M.I. Šantek, D. Komes, S. Novak, B. Šantek, Bioethanol production from renewable raw materials and its separation and purification: A review, Food Technol Biotechnol 56 (2018) 289. https://doi.org/10.17113/FTB.56.03.18.5546.
[19] Ó.J. Sánchez, C.A. Cardona, Trends in biotechnological production of fuel ethanol from different feedstocks, Bioresour Technol 99 (2008) 5270–5295. https://doi.org/10.1016/J.BIORTECH.2007.11.013.
[20] M. Balat, Production of bioethanol from lignocellulosic materials via the biochemical pathway: A review, Energy Convers Manag 52 (2011) 858–875. https://doi.org/10.1016/J.ENCONMAN.2010.08.013.
[21] P. Binod, R. Sindhu, R.R. Singhania, S. Vikram, L. Devi, S. Nagalakshmi, N. Kurien, R.K. Sukumaran, A. Pandey, Bioethanol production from rice straw: An overview, Bioresour Technol 101 (2010) 4767–4774. https://doi.org/10.1016/J.BIORTECH.2009.10.079.
[22] F. Cherubini, A.H. Strømman, Principles of biorefining, Biofuels: Alternative Feedstocks and Conversion Processes (2011) 3–24. https://doi.org/10.1016/B978-0-12-385099-7.00001-2.
[23] J.A. Quintero, L.E. Rincón, C.A. Cardona, Production of bioethanol from agroindustrial residues as feedstocks, Biofuels: Alternative Feedstocks and Conversion Processes (2011) 251–285. https://doi.org/10.1016/B978-0-12-385099-7.00011-5.
[24] B.E. Dale, S. Kim, Biomass refining global impact—The biobased economy of the 21st century, Biorefineries-Industrial Processes and Products: Status Quo and Future Directions 1 (2008) 41–66. https://doi.org/10.1002/9783527619849.CH2.
[25] K.O. Olatunji, N.A. Ahmed, O. Ogunkunle, Optimization of biogas yield from lignocellulosic materials with different pretreatment methods: A review, Biotechnol Biofuels 14 (2021). https://doi.org/10.1186/s13068-021-02012-x.
[26] A. Devi, S. Bajar, P. Sihag, Z.U.D. Sheikh, A. Singh, J. Kaur, N.R. Bishnoi, D. Pant, A panoramic view of technological landscape for bioethanol production from various generations of feedstocks, Bioengineered 14 (2023) 81–112. https://doi.org/10.1080/21655979.2022.2095702.
[27] J. Baeyens, Q. Kang, L. Appels, R. Dewil, Y. Lv, T. Tan, Challenges and opportunities in improving the production of bio-ethanol, Prog Energy Combust Sci 47 (2015) 60–88. https://doi.org/10.1016/J.PECS.2014.10.003.
[28] T.T.A. Tran, T.K.P. Le, T.P. Mai, D.Q. Nguyen, T.T.A. Tran, T.K.P. Le, T.P. Mai, D.Q. Nguyen, Bioethanol production from lignocellulosic biomass, Xiandai Huagong/Modern Chemical Industry 31 (2019) 40–44. https://doi.org/10.5772/INTECHOPEN.86437.
[29] Ó.J. Sánchez, C.A. Cardona, Trends in biotechnological production of fuel ethanol from different feedstocks, Bioresour Technol 99 (2008) 5270–5295. https://doi.org/10.1016/J.BIORTECH.2007.11.013.

[30] M.I. Šantek, E. Miškulin, S. Beluhan, B. Šantek, New Trends in the ethanol production as a biofuel, Kemija u Industriji: Časopis Kemičara i Kemijskih Inženjera Hrvatske 65 (2016) 25–38. https://doi.org/10.15255/KUI.2014.032.
[31] N.V. Narendranath, K.C. Thomas, W.M. Ingledew, Urea hydrogen peroxide reduces the numbers of Lactobacilli, nourishes yeast, and leaves no residues in the ethanol fermentation, Appl Environ Microbiol 66 (2000) 4187–4192. https://doi.org/10.1128/AEM.66.10.4187-4192.2000/ASSET/1562FA49-8311-44B7-96AF-962D09F02621/ASSETS/GRAPHIC/AM1000561005.JPEG.
[32] Ethanol Explained—Use of Ethanol—U.S. Energy Information Administration (EIA) (n.d.). https://www.eia.gov/energyexplained/biofuels/ethanol-use.php (accessed August 2, 2024).
[33] L.S. Khuong, H.H. Masjuki, N.W.M. Zulkifli, E.N. Mohamad, M.A. Kalam, A. Alabdulkarem, A. Arslan, M.H. Mosarof, A.Z. Syahir, M. Jamshaid, Effect of gasoline–bioethanol blends on the properties and lubrication characteristics of commercial engine oil, RSC Adv 7 (2017) 15005–15019. https://doi.org/10.1039/C7RA00357A.
[34] Y. Wang, G. Li, Z. Liu, P. Cui, Z. Zhu, S. Yang, Techno-economic analysis of biomass-to-hydrogen process in comparison with coal-to-hydrogen process, Energy 185 (2019) 1063–1075. https://doi.org/10.1016/J.ENERGY.2019.07.119.
[35] S. Sharma, S.K. Ghoshal, Hydrogen the future transportation fuel: From production to applications, Renew Sustain Energy Rev 43 (2015) 1151–1158. https://doi.org/10.1016/J.RSER.2014.11.093.
[36] R. Javaid, Catalytic hydrogen production, storage and application, Catalysts 11 (2021) 836. https://doi.org/10.3390/CATAL11070836.
[37] J.M. Sánchez, M.M. Barreiro, M. Maroño, Bench-scale study of separation of hydrogen from gasification gases using a palladium-based membrane reactor, Fuel 116 (2014) 894–903. https://doi.org/10.1016/J.FUEL.2013.02.051.
[38] C.A. Pereira, P.M. Coelho, J.F. Fernandes, M.H. Gomes, Study of an energy mix for the production of hydrogen, Int J Hydrogen Energy 42 (2017) 1375–1382. https://doi.org/10.1016/J.IJHYDENE.2016.07.182.
[39] T. Lepage, M. Kammoun, Q. Schmetz, A. Richel, Biomass-to-hydrogen: A review of main routes production, processes evaluation and techno-economical assessment, Biomass Bioenergy 144 (2021) 105920. https://doi.org/10.1016/J.BIOMBIOE.2020.105920.
[40] U.Y. Qazi, Future of hydrogen as an alternative fuel for next-generation industrial applications; challenges and expected opportunities, Energies 15 (2022) 4741. https://doi.org/10.3390/EN15134741.
[41] E.M. Barampouti, S. Mai, D. Malamis, K. Moustakas, M. Loizidou, Liquid biofuels from the organic fraction of municipal solid waste: A review, Renew Sustain Energy Rev 110 (2019) 298–314. https://doi.org/10.1016/J.RSER.2019.04.005.
[42] A. Singh, S. Sevda, I.M. Abu Reesh, K. Vanbroekhoven, D. Rathore, D. Pant, Biohydrogen Production from lignocellulosic biomass: Technology and sustainability, Energies 8 (2015) 13062–13080. https://doi.org/10.3390/EN81112357.
[43] V. Stenberg, M. Rydén, T. Mattisson, A. Lyngfelt, Exploring novel hydrogen production processes by integration of steam methane reforming with chemical-looping combustion (CLC-SMR) and oxygen carrier aided combustion (OCAC-SMR), Int J Greenhouse Gas Control 74 (2018) 28–39. https://doi.org/10.1016/J.IJGGC.2018.01.008.
[44] C. Xu, S. Chen, A. Soomro, Z. Sun, W. Xiang, Hydrogen rich syngas production from biomass gasification using synthesized Fe/CaO active catalysts, J Energy Inst 91 (2018) 805–816. https://doi.org/10.1016/J.JOEI.2017.10.014.
[45] C. Srinivasakannan, N. Balasubramanian, Variations in the design of dual fluidized bed gasifiers and the quality of syngas from biomass, Energy Sources Part A Recovery Util Environ Eff 33 (2010) 349–359. https://doi.org/10.1080/15567030902967835.

[46] P. Parthasarathy, K.S. Narayanan, Hydrogen production from steam gasification of biomass: Influence of process parameters on hydrogen yield—A review, Renew Energy 66 (2014) 570–579. https://doi.org/10.1016/J.RENENE.2013.12.025.
[47] A.B.H. Trabelsi, A. Ghrib, K. Zaafouri, A. Friaa, A. Ouerghi, S. Naoui, H. Belayouni, Hydrogen-Rich Syngas Production from Gasification and Pyrolysis of Solar Dried Sewage Sludge: Experimental and Modeling Investigations, Biomed Res Int (2017) 7831470. https://doi.org/10.1155/2017/7831470.
[48] A. Bakhtyari, M.A. Makarem, M.R. Rahimpour, Hydrogen production through pyrolysis, Fuel Cells and Hydrogen Production (2019) 947–973. https://doi.org/10.1007/978-1-4939-7789-5_956.
[49] M.G. Eloffy, A.M. Elgarahy, A.N. Saber, A. Hammad, D.M. El-Sherif, M. Shehata, A. Mohsen, K.Z. Elwakeel, Biomass-to-sustainable biohydrogen: Insights into the production routes, and technical challenges, Chem Eng J Adv 12 (2022) 100410. https://doi.org/10.1016/J.CEJA.2022.100410.
[50] D.B. Levin, L. Pitt, M. Love, Biohydrogen production: Prospects and limitations to practical application, Int J Hydrogen Energy 29 (2004) 173–185. https://doi.org/10.1016/S0360-3199(03)00094-6.
[51] C.N.C. Hitam, A.A. Jalil, A review on biohydrogen production through photo-fermentation of lignocellulosic biomass, Biomass Convers Biorefin 13 (2023) 8465–8483. https://doi.org/10.1007/S13399-020-01140-Y/METRICS.
[52] H. Bothe, O. Schmitz, M.G. Yates, W.E. Newton, Nitrogen fixation and hydrogen metabolism in cyanobacteria, Microbiol Mol Biol Rev 74 (2010) 529–551. https://doi.org/10.1128/MMBR.00033-10/ASSET/6A89F896-E2C7-4CB2-B80B-3DABFEBFDB6F/ASSETS/GRAPHIC/ZMR9990922580010.JPEG.
[53] A. Kadier, A.K. Chaurasia, S.M. Sapuan, R.A. Ilyas, P.C. Ma, K.F.S. Alabbosh, P.K. Rai, W. Logroño, A.A. Hamid, H.A. Hasan, Essential factors for performance improvement and the implementation of microbial electrolysis cells (MECs), Bioelectrochemical Systems: Vol. 1 Principles and Processes (2020) 139–168. https://doi.org/10.1007/978-981-15-6872-5_7.
[54] J.L. Varanasi, R. Veerubhotla, S. Pandit, D. Das, Biohydrogen production using microbial electrolysis cell: Recent advances and future prospects, Biomass, Biofuels, Biochemicals: Microbial Electrochemical Technology: Sustainable Platform for Fuels, Chemicals and Remediation (2019) 843–869. https://doi.org/10.1016/B978-0-444-64052-9.00035-2.
[55] S. Bajracharya, M. Sharma, G. Mohanakrishna, X. Dominguez Benneton, D.P.B.T.B. Strik, P.M. Sarma, D. Pant, An overview on emerging bioelectrochemical systems (BESs): Technology for sustainable electricity, waste remediation, resource recovery, chemical production and beyond, Renew Energy 98 (2016) 153–170. https://doi.org/10.1016/J.RENENE.2016.03.002.
[56] J. Sentanuhady, W.H. Hasan, M.A. Muflikhun, Recent progress on the implementation of renewable biodiesel fuel for automotive and power plants: Raw materials perspective, Adv Mater Sci Eng (2022) 5452942. https://doi.org/10.1155/2022/5452942.
[57] R.A. Lee, J.M. Lavoie, From first- to third-generation biofuels: Challenges of producing a commodity from a biomass of increasing complexity, Animal Front 3 (2013) 6–11. https://doi.org/10.2527/AF.2013-0010.
[58] K. Velusamy, J. Devanand, P. Senthil Kumar, K. Soundarajan, V. Sivasubramanian, J. Sindhu, D.V.N. Vo, A review on nano-catalysts and biochar-based catalysts for biofuel production, Fuel 306 (2021) 121632. https://doi.org/10.1016/J.FUEL.2021.121632.
[59] J. Maroušek, O. Strunecký, V. Bartoš, M. Vochozka, Revisiting competitiveness of hydrogen and algae biodiesel, Fuel 328 (2022) 125317. https://doi.org/10.1016/J.FUEL.2022.125317.

[60] S.P. Jeevan Kumar, L.K.S. Srinivas Gujjala, A. Dash, B. Talukdar, R. Banerjee, Biodiesel production from lignocellulosic biomass using oleaginous microbes, Lignocellulosic Production and Industrial Applications (2017) 65–92. https://doi.org/10.1002/9781119323686.CH4.
[61] E. Betiku, S.O. Ajala, Modeling and optimization of Thevetia peruviana (yellow oleander) oil biodiesel synthesis via Musa paradisiacal (plantain) peels as heterogeneous base catalyst: A case of artificial neural network vs. response surface methodology, Ind Crops Prod 53 (2014) 314–322. https://doi.org/10.1016/J.INDCROP.2013.12.046.
[62] E. Paone, A. Tursi, Production of biodiesel from biomass, Advances in Bioenergy and Microfluidic Applications (2021) 165–192. https://doi.org/10.1016/B978-0-12-821601-9.00006-6.
[63] H. Fukuda, A. Kondo, H. Noda, Biodiesel fuel production by transesterification of oils, J Biosci Bioeng 92 (2001) 405–416. https://doi.org/10.1016/S1389-1723(01)80288-7.
[64] H. Wang, X. Peng, H. Zhang, S. Yang, H. Li, Microorganisms-promoted biodiesel production from biomass: A review, Energy Convers Manag X 12 (2021) 100137. https://doi.org/10.1016/J.ECMX.2021.100137.
[65] N. Kanthavelkumaran, P. Dr, Seenikannan, Recent trends and applications of bio diesel, Int J Eng Res Appl (IJERA) 2 (n.d.) 197–203. www.ijera.com (accessed August 3, 2024).
[66] T. Somorin, J. Gitau, D. Agbefu, S. Gebrezgabher, Biomass Briquetting: A Training Module for Trainers and Practitioners (2023). www.cgiar.org/funders (accessed August 3, 2024).
[67] H.M.P. Marreiro, R.S. Peruchi, R.M.B.P. Lopes, S.L.F. Andersen, S.A. Eliziário, P.R. Junior, Empirical studies on biomass briquette production: A literature review, Energies 14 (2021) 8320. https://doi.org/10.3390/EN14248320.
[68] T. Wongsiriamnuay, N. Tippayawong, Effect of densification parameters on the properties of maize residue pellets, Biosyst Eng 139 (2015) 111–120. https://doi.org/10.1016/J.BIOSYSTEMSENG.2015.08.009.
[69] S. Khlifi, M. Lajili, S. Belghith, S. Mezlini, F. Tabet, M. Jeguirim, Briquettes production from olive mill waste under optimal temperature and pressure conditions: Physico-chemical and mechanical characterizations, Energies 13 (2020) 1214. https://doi.org/10.3390/EN13051214.
[70] V. Panwar, B. Prasad, K.L. Wasewar, Biomass residue briquetting and characterization, J Energy Eng 137 (2010) 108–114. https://doi.org/10.1061/(ASCE)EY.1943-7897.0000040.
[71] S.Y. Kpalo, M.F. Zainuddin, L.A. Manaf, A.M. Roslan, A review of technical and economic aspects of biomass briquetting, Sustainability (Switzerland) 12 (2020). https://doi.org/10.3390/SU12114609.
[72] D.K. Okot, P.E. Bilsborrow, A.N. Phan, Effects of operating parameters on maize COB briquette quality, Biomass Bioenergy 112 (2018) 61–72. https://doi.org/10.1016/J.BIOMBIOE.2018.02.015.
[73] H. Gilvari, W. de Jong, D.L. Schott, Quality parameters relevant for densification of bio-materials: Measuring methods and affecting factors—A review, Biomass Bioenergy 120 (2019) 117–134. https://doi.org/10.1016/J.BIOMBIOE.2018.11.013.
[74] F.F. Felfli, J.M. Mesa P, J.D. Rocha, D. Filippetto, C.A. Luengo, W.A. Pippo, Biomass briquetting and its perspectives in Brazil, Biomass Bioenergy 35 (2011) 236–242. https://doi.org/10.1016/J.BIOMBIOE.2010.08.011.
[75] G. Wang, Y. Dai, H. Yang, Q. Xiong, K. Wang, J. Zhou, Y. Li, S. Wang, A review of recent advances in biomass pyrolysis, Energy Fuels 34 (2020) 15557–15578. https://doi.org/10.1021/ACS.ENERGYFUELS.0C03107/ASSET/IMAGES/MEDIUM/EF0C03107_0014.GIF.

[76] B. Hu, J. Liu, W. luan Xie, Y. Li, Q. Lu, Biofuels production using pyrolysis techniques, Advances in Biofuels Production, Optimization and Applications (2024) 103–125. https://doi.org/10.1016/B978-0-323-95076-3.00010-7.
[77] K.O. Olatunji, D.M. Madyira, A.M. Kożuch, Biomass resources assessment and pretreatment technologies, Renewable Energy Technologies (2026) 3–31. https://doi.org/10.1016/B978-0-443-33771-0.00001-0.
[78] M. Balat, An overview of the properties and applications of biomass pyrolysis oils, Energy Sources Part A Recovery, Util Environ Eff 33 (2011) 674–689. https://doi.org/10.1080/15567030903228914.
[79] J. Cai, N. Lin, Y. Li, J. Xue, F. Li, L. Wei, M. Yu, X. Zha, W. Li, Research on the application of catalytic materials in biomass pyrolysis, J Anal Appl Pyrolysis 177 (2024) 106321. https://doi.org/10.1016/J.JAAP.2023.106321.
[80] A.K. Vuppaladadiyam, S.S.V. Vuppaladadiyam, A. Awasthi, A. Sahoo, S. Rehman, K.K. Pant, S. Murugavelh, Q. Huang, E. Anthony, P. Fennel, S. Bhattacharya, S.Y. Leu, Biomass pyrolysis: A review on recent advancements and green hydrogen production, Bioresour Technol 364 (2022). https://doi.org/10.1016/J.BIORTECH.2022.128087.
[81] B. Van De Beld, E. Holle, J. Florijn, The use of pyrolysis oil and pyrolysis oil derived fuels in diesel engines for CHP applications, Appl Energy 102 (2013) 190–197. https://doi.org/10.1016/J.APENERGY.2012.05.047.
[82] T. Kan, V. Strezov, T. Evans, J. He, R. Kumar, Q. Lu, Catalytic pyrolysis of lignocellulosic biomass: A review of variations in process factors and system structure, Renew Sustain Energy Rev 134 (2020). https://doi.org/10.1016/j.rser.2020.110305.
[83] W. Wang, Y. Gu, C. Zhou, C. Hu, Current challenges and perspectives for the catalytic pyrolysis of lignocellulosic biomass to high-value products, Catalysts 12 (2022) 1524. https://doi.org/10.3390/CATAL12121524.
[84] A. Molino, V. Larocca, S. Chianese, D. Musmarra, Biofuels production by biomass gasification: A review, Energies 11 (2018) 811. https://doi.org/10.3390/EN11040811.
[85] V.S. Sikarwar, M. Zhao, P.S. Fennell, N. Shah, E.J. Anthony, Progress in biofuel production from gasification, Prog Energy Combust Sci 61 (2017) 189–248. https://doi.org/10.1016/J.PECS.2017.04.001.
[86] A.A. Ahmad, N.A. Zawawi, F.H. Kasim, A. Inayat, A. Khasri, Assessing the gasification performance of biomass: A review on biomass gasification process conditions, optimization and economic evaluation, Renew Sustain Energy Rev 53 (2016) 1333–1347. https://doi.org/10.1016/J.RSER.2015.09.030.
[87] B. Matas Güell, J. Sandquist, L. Sørum, Gasification of biomass to second generation biofuels: A review, J Energy Resour Technol 135 (2013). https://doi.org/10.1115/1.4007660.
[88] G. Enaime, A. Baçaoui, A. Yaacoubi, M. Lübken, Biochar for wastewater treatment—Conversion technologies and applications, Appl Sci 10 (2020) 3492. https://doi.org/10.3390/APP10103492.
[89] S. Wijitkosum, Biochar derived from agricultural wastes and wood residues for sustainable agricultural and environmental applications, Int Soil Water Conserv Res 10 (2022) 335–341. https://doi.org/10.1016/J.ISWCR.2021.09.006.
[90] J.A. Antonangelo, X. Sun, H. Zhang, The roles of co-composted biochar (COMBI) in improving soil quality, crop productivity, and toxic metal amelioration, J Environ Manage 277 (2021) 111443. https://doi.org/10.1016/J.JENVMAN.2020.111443.
[91] E.D. Frank, A. Elgowainy, K. Reddi, A. Bafana, Life-cycle analysis of greenhouse gas emissions from hydrogen delivery: A cost-guided analysis, Int J Hydrogen Energy 46 (2021) 22670–22683. https://doi.org/10.1016/J.IJHYDENE.2021.04.078.
[92] D. Gerrior, K. Delsoz Bahri, A. Kermanshahi-pour, M.J. Eckelman, S.K. Brar, Life cycle assessment and techno-economic analysis of a novel closed loop corn ethanol biorefinery, Sustain Prod Consum 30 (2022) 359–376. https://doi.org/10.1016/J.SPC.2021.12.007.

[93] C. Rewlay-ngoen, S. Papong, R. Onbhuddha, B. Thanomnim, Evaluation of the environmental performance of bioethanol from cassava pulp using life cycle assessment, J Clean Prod 284 (2021) 124741. https://doi.org/10.1016/J.JCLEPRO.2020.124741.
[94] P.J. De Wild, W.J.J. Huijgen, Organosolv Fractionation of Lignocellulosic Biomass for an Integrated Biorefinery Durzame Technologie (2015).
[95] S.I. Mussatto, G.M. Dragone, Biomass pretreatment, biorefineries, and potential products for a bioeconomy development, Biomass Fractionation Technologies for a Lignocellulosic Feedstock Based Biorefinery (2016) 1–22. https://doi.org/10.1016/B978-0-12-802323-5.00001-3.

PART II

Lignocellulose biomass pretreatment

3 Biological pretreatment

3.1 INTRODUCTION

Biological pretreatment is the process of using the ligninolytic properties of microorganisms and enzymes to lower the recalcitrant structure of lignocellulose feedstocks and improve their accessibility to hydrolytic enzymes during bioenergy production, as shown in Figure 3.1. Typically, this process involves growing microorganisms directly within the feedstocks or through the use of enzyme cocktails [1]. The technique is performed at low temperatures and pressures (near ambient), therefore requiring less energy and chemical inputs with lower inhibitory compounds. Compared with conventional physical or chemical pretreatments, this technique is mostly environmentally benign. The process is environmentally friendly because it does not require any chemicals, eliminating the need for chemical recovery/recycling, and it does not release any toxic or hazardous materials into the environment. In this technique, different microbes are applied to degrade or break down the lignocellulose structural arrangement by producing hydrolytic enzymes (hydrolases) and ligninolytic enzymes that depolymerize the lignin. The cell wall arrangement is then opened for subsequent hydrolysis of the lignocellulose feedstock. Hemicellulose and cellulose are hydrolyzed into monomeric sugars by cellulolytic and hemicellulolytic microorganisms [2]. When lignocellulose biomass degradation and the subsequent fermentation process occur simultaneously, it releases biofuels such as methane, ethanol, and hydrogen, as well as other bioproducts, including organic acids, acetates, enzymes, and lactate [3]. Different microorganisms such as *Clostridium, Trichoderma reesei, Thermomonospora sp.*, *Trichoderma viride*, *Streptomyces sp.*, *Aspergillus niger.*, *Phanerochaete chrysosporium,* etc., have been identified to have the strength to hydrolyze lignocellulose feedstocks [2]. Biological pretreatments can be utilized for other biorefinery applications beyond feedstock pretreatment. For instance, the enzymes produced can be separated and utilized for other purposes. Organic compounds and lignin derivatives can be generated through this process, which can also serve as sources of chemicals and add value to the process [4]. One of the limitations of this technique is the inability to convert the non-fermentable portions of the lignocellulose feedstock into value-added products, which reduces the biorefinery's efficiency. Another challenge of the technique is the longer incubation period and lower overall efficiency (sugar yield), which necessitates enhancements to make the technique competitive with other thermochemical pretreatment techniques [5]. Biological pretreatments are classified based on the type of microorganism applied: bacteria, enzymes, fungi, ensiling, microaerobic, and bioaugmentation pretreatments.

DOI: 10.1201/9781003633815-6

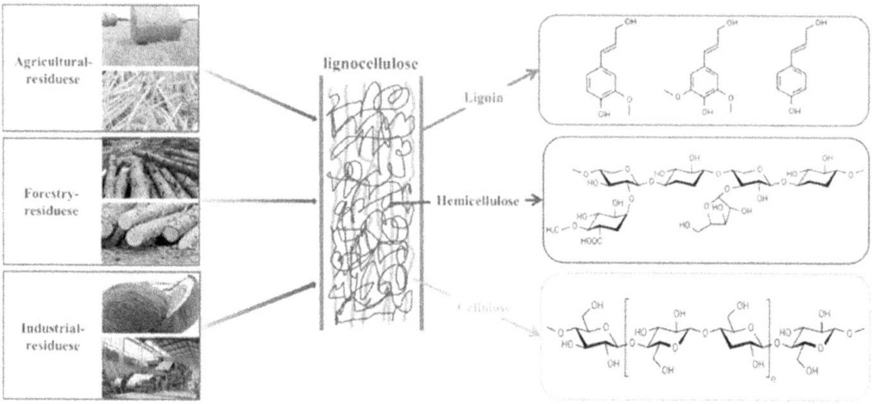

FIGURE 3.1 Biological pretreatment of lignocellulose biomass [6].

3.2 BACTERIA

Bacteria pretreatment is a biological pretreatment technique that utilizes bacteria that produce enzymes to alter the lignin and hemicellulose arrangement of lignocellulose materials, thereby making cellulose accessible for biorefining. There are different bacteria that release varying types of enzymes that can degrade biomass. They are classified as ligninolytic, hemicellulolytic, and cellulolytic enzymes [2]. Ligninolytic enzymes are further classified as lignin peroxidase, versatile peroxidase, laccase, manganese peroxidase, and cellobiose dehydrogenase, with examples of bacterial species of *Bacillus subtilis, Raoultella sp., Azospirillum lipoferum, Cupriavidus basilensis, and Prevotella sp*, respectively. Hemicellulolytic enzymes comprise β-glucosidase, α-galactosidase, endoglucanase, Xylanases, acetyl esterase, β-xylosidase, and mannanase, with bacterial species such as *Bacteroides succinogenes*, *Fibrobacter succinogenes*, *Prevotella bryantii*, *Bacillus sp.*, *Ruminococcus albus*, *Pseudobutyrivibrio xylanivorans*, and *Ruminococcus flavefaciens*, respectively, as examples. Cellulolytic enzymes are further classified into endoglucanase, cellobiohydrolase, β-glucosidase, and glycosyltransferases, with Clostridium sp., Bacillus sp., Pedobacter sp., and Cellulomonas sp. as examples [2–4]. During the pretreatment of bacteria, enzymatic hydrolysis, and fermentation of lignocellulose biomass in biofuel production, selecting the most effective and economical bacteria is a crucial stage. Compared to lignin, hemicellulose and cellulose can be degraded easily. An extensive investigation has been conducted into the use of cellulolytic bacteria, such as *Thermomospora fusca* and *Cellulomonas fimi*, for cellulase generation. The study has shown that cellulolytic bacteria, such as *Paenibacillus campinasensis*, possess the ability to survive under harsh conditions and exhibit excellent properties for lignocellulose biomass pretreatment [7,8]. Despite several cellulolytic bacteria, such as *Bacteroides cellulosolvens* and *Clostridium thermocellum*, that exhibit high cellulase performance, the challenge lies in their inability to produce sufficient enzymatic concentrations [9]. However, bacteria such as *Zymomonas mobilis*, a notable cellulolytic bacterium, can be utilized for fermenting glucose, sucrose, and fructose to improve ethanol production [2]. Another important bacterium that has been observed for cellulase production and higher performance in cellulosic feedstock degradation is the gram-positive *Bacillus* strain, *Firmicutes*, as well as gram-negative strains *Rahnella, Pseudomonas*, and *Buttiauxella* [10]. Bacterial species like *Bacillus subtilis* and *Azospirillum lipoferum* have been reported to release bacterial laccases that can depolymerize lignin [11]. Studies on the microbial depolymerization of lignin using bacteria are limited compared to those using fungi. However, there has been a renewed interest in lignin depolymerization using bacteria recently due to the discovery of bacterial laccases, peroxidases, and β-etherases that can efficiently delignify lignin [12–15].

Bacterial strains, such as Bacillus, possess sufficient genes encoding glycoside hydrolases and glycosyltransferases that can be utilized to degrade cellulose. This strain was applied to inoculate rice straw, an improvement in the microenvironment, and an increase in the gathering of bacterial communities to break down the lignocellulose feedstock [16]. The application of marine bacteria for lignocellulose pretreatment using nine strains with endoglucanase activity from *Mesodesma donacium* and *Mytilus chinensis*. The selected strains, including *Chryseobacterium, Pseudomonas, Raoultella,* and *Aeromonas*, were observed to be able to degrade the filter paper. However, it was observed that the *Aeromonas bivalvium* strain MC23 and *Raoultella ornithinolytica* strains MC3 and MA5 produced more significant impacts, with 3.51, 4.23, and 5.26% degradation, respectively. Utilization of microalgae pretreatment with MC3 and MA5 strains shows a significant cell wall damage at 30 °C, which supports subsequent bioconversion, and it requires lower energy, which is one of its major merits [17]. The use of *Ochrobactrum oryzae* BMP03 is extracted from alkali lignin medium, which grows at a pH of 6 and an optimal temperature of 30 °C. At the optimal conditions, the strain produces peroxidase and laccase to break down lignin and attains the logarithmic stage within the sixth and ninth days [18]. Using *Arthrobacter sp.* C2, a low-temperature-tolerant bacterium strain used for pretreating sodium lignosulfonate feedstock, produces MnP and Lip to break down lignin. It was observed that the strain achieved an optimum degradation rate of 39.2% at 15 °C and a pH of 3–10. The possible breakdown involves hydroxyl group oxidation, carbon–oxygen link cleavage, methoxy group demethylation, and carboxyl group decarboxylation [19]. In general, it was observed that different strains of bacteria produce varying degrees of influence on lignocellulose feedstocks. Bacteria can adjust quickly to an adverse environment than fungi, which makes them more applicable in lignocellulose pretreatment.

3.3 ENZYMES

In this pretreatment method, enzymes are used to alter the microstructural arrangement of lignocellulosic biomass, thereby improving its accessibility during the subsequent biorefinery process. The primarily utilized enzymes include manganese peroxidase and β-lactamases, which are usually regarded as degrading enzymes for lignin, hemicellulose, and cellulose [20]. Manganese peroxidase and lactase enzymes assisted in removing fermentable sugars from the lignocellulose feedstocks, which enhances subsequent biofuel production. Glycosidases and proteases are enzymes commonly used for biosludge materials, as the feedstock is a composition of microbial biomass rich in complex carbohydrates and proteins. Biosludge feedstocks are enclosed in a gel-like matrix of extracellular polymeric materials that have various biopolymers such as lignin, proteins, carbohydrates, RNA, and DNA [21]. Enzyme pretreatment has the capacity to navigate some of the operational challenges of anaerobic digestion. This method is very efficient when the correct enzyme and optimal operating conditions, such as dosage, are used [22]. Compared to other biological pretreatments, such as microbial pretreatment, enzymatic pretreatment does not require an acclimatization stage to the feedstock, which implies that there is no process delay. Enzymes are biodegradable proteins that degrade themselves without generating new biomass. Biomass pretreatment using enzymes can be performed under mild operating conditions, and it can be conducted under varying environmental conditions, such as temperature and pH. The feasibility and cost have been major concerns for enzyme pretreatment due to the very long exposure time. In comparison to chemical, physical, and thermal pretreatment, enzymatic pretreatment requires lower energy since the enzymes operate at a mild temperature [23]. The impact of enzymatic pretreatment on lignocellulose biomass relies on the enzyme types and the morphological arrangement of the biomass treated [24,25].

Green macroalgae Ulva sp. was pretreated with Aspergillus fumigatus SL1, and it was observed that the cost of pretreatment is lower than that of chemical pretreatment [26]. Biosludge was pretreated with enzymes, and protease from *B. licheniformis* increased the biomethane by 26% after a 62-day retention period. Other enzymes experimented with during this study did not improve biogas yield, except for protease BCE-2078. The chemical oxygen demand (COD) of the enzyme was observed not to be completely digested into biogas; in some situations, the enzyme affects the inoculum negatively and reduces biogas release [21]. The laccase enzyme, produced from phenoloxidase that contains various copper ions, was able to catalyze the oxidation of phenols, aromatic thiols, and anilines in the feedstock. Therefore, the microbial development was improved, and the fermentation rate was increased during anaerobic digestion [27]. Manganese peroxidase enzyme generated from white-rot fungi was observed to catalyze the oxidation of Mn^{2+} to Mn^{3+} and increase the degradation of lignin linkages [28]. The cellulase enzyme produced from the *Trichoderma reesei* fungus has been noticed for its ability to break down the insoluble cellulose present in the vast majority of lignocellulose biomass [29]. This enzyme's activity catalyzes the biochemical process during anaerobic digestion, leading to improved lignocellulose degradation and a subsequent increase in biogas yield. Paper and pulp sludge was pretreated with endoglucanase and laccase, resulting in a 134% enhancement in methane yield [30]. Sewage sludge pretreatment with 10% endogenous hydrolases (protease from *A. hydrophila* and amylases from *B. subtilis*) and digested for 11 days was observed to enhance biogas production by 23.1% when both enzymes were used in combination [31]. In a similar study, Bonilla *et al.* investigated the use of self-produced and commercial proteases in paper biosludge pretreatment prior to testing for biomethane potential. It was discovered that the self-produced protease BCE-2078 did not show any significant improvement in biogas yield; however, commercial proteases (*Bacillus licheniformis*) increased the biogas yield by 26% after a 62-day retention period [21].

The performance of the enzymatic pretreatment technique depends on various factors, including the strain, temperature, pH, microstructural arrangement of the feedstock, and exposure time. Recent investigations into enzyme pretreatment have garnered more attention due to its ability to enhance bioenergy production through the biorefinery of lignocellulose biomass. Several studies have demonstrated that enzymatic pretreatment is a promising technique for enhancing biofuel production from lignocellulose feedstocks. Over 40 enzymes have been examined during the enzymatic pretreatment of lignocellulose feedstock, and cellulase is the most considered, followed by β-glucosidase and xylanase [32]. There is no extensive study on most of these enzymes at the laboratory scale before considering them at the industrial scale. There are several advantages associated with enzymatic pretreatment, including a lower environmental impact and reduced energy requirement compared to other conventional pretreatment methods. The major limitation of this method is the high cost of the enzyme and the longer exposure period [33]. Further studies are required to assess the influence of enzymatic treatment on improved biofuel production, particularly for some enzymes that have received less attention. Additionally, the contribution of COD digestion to enzyme actions and the influence of enzymatic activity on inactive enzymes need to be evaluated separately.

3.4 FUNGI

Fungi pretreatment of biomass is a biological technique that utilizes fungi to break down lignin, thereby improving the accessibility of biomass for subsequent processes, such as anaerobic digestion or enzymatic hydrolysis. It is an environmentally benign and energy-efficient substitute for conventional thermal or chemical pretreatment methods [34]. This method is applied at 25 to 30 °C and atmospheric pressure, with low water requirements and no chemical addition, eliminating the need for recovery or disposal. Fungi are popular microorganisms because of their interactive influence on decaying materials through their enzyme. Lignocellulose biomass is degraded through excretion because fungi have a limited diffusion capacity, and the level of colonization is crucial for efficient biomass degradation, determining the success of pretreatment [35]. Compounds with a complex mixture are produced during fungal colonization, and these compounds can break down the lignocellulose biomass. Different fungal species employ distinct strategies and exhibit varying capacities to break down biomass [36]. Wood-rotting fungi such as brown-, white-, and soft-rot fungi are generally applied during pretreatment because of the capacity to alter the morphological arrangement of lignocellulose feedstock [37]. Brown-rot fungi primarily alter hemicellulose and cellulose with limited impact on lignin, which does not support the generation of fermentable sugars [4]. There are certain soft-rot fungi that can break down lignin, whereas the majority can partially alter it but degrade the structural carbohydrates significantly [38]. White-rot fungi can totally mineralize lignin during aerobic conditions and, in some instances, do so discriminately than cellulose, and this makes white-rot fungi preferable [39]. Fungi are widely available in nature and generate different ligninolytic, cellulolytic, and hemicellulolytic enzymes [2]. Fungal species from basidiomycetes, which include white-rot fungi (e.g., P.chrysosporium, Schizophyllum sp.), ascomycetes (e.g., Aspergillus sp., Trichoderma reesei, Penicillium sp.), brown-rot fungi (e.g., *Fomitopsis palustris*), and some anaerobic strains (*Orpinomyces sp.*) are examples of lignocellulolytic fungi [40,41].

A study on *T. reesei* indicated that it releases a significant amount of β-glucosidase and xylanases, which are high in cellulase activities [42]. Likewise, a detailed investigation of the soil fungus Trichoderma longibrachiatum revealed that it is one of the most promising strains in terms of crystalline solubility due to its ability to produce three distinct types of cellulases: β-glucosidases (e.g., cellobioases), endoglucanases (e.g., carboxymethyl cellulases), and exoglucanases (e.g., *cellobiohydrolases*). These varying cellulases and biomass have complex relationships that function in a synergistic manner during hydrolysis [43,44]. Due to its complex arrangement, it is challenging to determine the most effective fungal strain for lignin. White-rot fungi, such as basidiomycetes, are recognized for their ability to disintegrate lignin and are considered conventional lignin-degrading microorganisms. They produce a wide range of ligninolytic enzymes, including manganese peroxidases, laccases, and lignin peroxidases, which can break down and mineralize lignin [45,46]. Four strains of fungi: *P. ostreatus* (strain MES 03449), *L. edodes* (strain MES 11910), *G. lucidum* (strain MES 02102), and *P.eryngii* (strain MES 03757) were used to pretreat wheat straw, wood chips, miscanthus, and rice straw. It was observed that *L. edodes* had the highest selective degrading strains of the considered strains and had a significant influence on wood chips, miscanthus, and wheat straw. The nutrient concentration findings indicate an increase in cellulose content; however, the mass balance suggests cellulose degradation. The most significant influence was observed on wood chips, wheat straw, and miscanthus during the 12-week exposure period. The growth of *L. edodes* development was positively related to lignin breakdown and total solids, and it varied with feedstock type [35]. Lignocellulose feedstock was pretreated with fungi using strains of *Cyclocybe aegerita* (MUT 5639), *Coprinopsis cinerea* (MUT 6385), and *Cephalotrichum stemonitis* (MUT 6326) to enhance the hydrolysis stage during anaerobic digestion. It was observed that the induced fungal pretreatment alters the feedstock cell wall with different impacts on each strain. After 20 days, it was observed that C. stemonitis produced the optimum cellulose, hemicellulose, and lignin reduction (8.2, 59.3, and 9.6%, respectively), and the highest total biogas and methane increases of 182 and 214%, respectively, compared to the untreated substrate [47]. Fungal pretreatment of non-hardwood, softwood, and sterile miscanthus with colonized substrate as inoculum was observed to increase the glucose yield by 4.5, 3, and 2-fold, and xylose produced by 10, 7, and 9-fold, respectively, after 14 days of incubation. A higher yield was also reported after 28 days when the feedstocks were sterile before fungal pretreatment. When non-sterile corn stover was pretreated with fungi, no significant improvement was observed. There was evidence of fungal colonization and significant cell wall degradability on the pretreated feedstocks, especially hardwood and miscanthus. It was also noted that fungal pretreatment had a more significant influence on woods than on herbaceous substrates, as evidenced by the alteration of morphological arrangement and substrate composition [48].

The efficiency of fungal pretreatment depends on the feedstock morphological arrangement, fungal strain, and culture conditions [49]. Fungi pretreatment typically requires low temperatures and chemicals, and it is unlikely to generate inhibitory compounds; therefore, detoxification or post-pretreatment cleaning/washing is not necessary. Furthermore, the method is efficient for substrates with bigger feedstock particles compared to some of the traditional pretreatment methods [5]. Nonetheless, this pretreatment method has some limitations compared to the conventional pretreatment techniques. This method requires low energy with little or no chemicals, but the exposure treatment time is longer, which is a challenge for potential industrial applications. The need for carbon sources for fungi development, which lowers the methane production, is another challenge [50]. The selection of fungal strains that are most efficient for lignin degradation and cellulose recovery simultaneously remains a challenge, and no breakthrough has been achieved yet. The requirement for substrate sterilization and lower sugar releases is another major challenge facing fungi pretreatment [4].

3.5 ENSILING

Ensiling is a technology that has been widely used in recent centuries, primarily for producing feed for livestock. It is a conventional technique that offers a bright means for storing moist biomass prior to biofuel production. It has minimal storage loss and is not affected by weather conditions compared to hay and open-air storage [51]. Lignocellulose biomasses are seasonal and require storage to maintain a feedstock supply for bioenergy production during off-season periods. This technique is a biochemical process where anaerobic microbes degrade water-soluble carbohydrates (like sucrose) into organic acids such as acetic and lactic acids. This mostly resulted in biological acidification and stable storage of biomass. Acid formation during ensiling lowers the pH values to ranges that inhibit the activity of microorganisms and preserve the organic matter [52]. However, about 18–35% of the organic matter can be lost during this technique if the organic acid levels that inhibit microbial activity are not achieved. These losses result from butyrate fermentation (the generation of CO_2 and H_2) caused by undesirable bacteria or methanogenic processes (CH_4 release) [53]. For effective biological acidification and further biofuel production, it is crucial to control storage conditions from the initial stage, taking into account the characteristics of the biomass, such as moisture, microbial control, and sugar amendment [54]. This technique is grouped into four stages: aerobic, anaerobic, stable, and discharge stage [51]. Each stages have some varying chemical and physical changes. During the aerobic stage, respiration is dominant until nearly all the available oxygen is consumed. At the next stage, anaerobic, facultative anaerobic, and anaerobic microbes, such as yeasts, lactic acid bacteria, and clostridia, are dominant. The quality of the biomass is determined at this stage; if lactic acid bacteria are the most dominant, a large quantity of lactic acid will be released, accompanied by a low pH, which can hinder the development of other microbes. Microbes' activities and pH levels are low at the stable stage, but the microorganism's metabolism is high at the final stage due to exposure to air, and organic carbon is lost at this stage [51].

Ensiling pretreatment has been experimented with different lignocellulose biomasses such as energy grass, sugarcane waste, crop straws, and whole-plant corn [52,55,56]. It has been observed that the technique can result in a loss of organic matter ranging from 0.1 to 20% during pretreatment [57]. Biomass ensiling alters the morphological structure (including depolymerization and crystallinity) of cellulose and hemicellulose but has a marginal impact on the lignin portion [54]. The influence of ensiling can be positive or negative on the same biomass, depending on the different capacities of biomass to accommodate organic mass degradation during the process. The efficiency of the process is directly proportional to the properties of the lignocellulose biomass [58]. The interaction between methane release and the chemical composition of lignocellulose biomass after ensiling was investigated, and it was observed that lignin content was the only component that had a visible negative impact on methane production potential. Likewise, the lignocellulose content, morphological arrangement, including lignin cross-linkage with polysaccharides, and the properties of cellulose and hemicellulose are connected to the quality of silage [59]. The impacts of wheat straw and sugar beet co-ensiling on biomethane yield were investigated at both laboratory and pilot scales. It was observed that, at the end of 9 months, co-ensiling enhanced the biomethane yield by 19–34% at the laboratory scale and by 18–32% at the pilot scale after 6 months. The findings indicate a mass loss of approximately 0.1%. It was concluded that green and straw biomass co-ensiling as a biological pretreatment can improve biomethane generation [57]. Switchgrass pretreatment using ensiling has been reported to be effective in biogas production, as it minimizes organic matter loss and enhances the biogas release rate. The hemicellulose and cellulose content were reduced, while the available soluble contents and organic acids were improved, and the bacterial community formation during anaerobic digestion was altered to increase methane generation [56]. The biomethane yield of ensiling-pretreated maize stover, which was ensiled for 60 days, was reported to increase by 16.5% during anaerobic digestion [55]. However, when wheat straw was ensiled 180 days prior to anaerobic digestion, it was observed that the biomethane yield was reduced by 12.3% compared to the untreated substrate [60]. Similarly, corn stalk ensiling for 30 days was observed to increase the volatile fatty acid content in the first 10 days, while the pH was reduced from 5.80 to less than 4.20, which subsequently inhibited methane yield, and the methane released was reduced by 9.22% compared to the control [61]. These findings imply that the efficiency of ensiling pretreatment and its role in biomethane production need further investigation. Additionally, there is a need to investigate the dramatic material and microbial transformations that occur during the stable ensiling phase and their impact on biogas generation and the microbial community in the digester. The influence of process parameters, such as carbohydrate and moisture contents, on the ensiling of biomass for different feedstocks needs to be further investigated.

3.6 MICROAEROBIC

Microaerobic/microaeration pretreatment is a process that introduces specific amounts of oxygen into the anaerobic digester to enhance the hydrolysis stage of lignocellulose digestion. This technique can be applied as a feedstock pretreatment or to adjust the microbial composition. Air/oxygen loading varies in studies during this process, depending on the aims [62]. The aim of this is to stimulate microbial development and performance at the hydrolysis stage, which is reported as the rate-limiting step. It has been reported as an effective pretreatment that disrupts the morphological arrangement of the feedstock with minimal cost and technical resources [63]. Microaerobic conditions have been employed for biogas upgrading during desulfurization; however, studies have also demonstrated their potential to enhance the hydrolysis, acidogenesis, and methanogenesis stages of anaerobic digestion. Microaerobic anaerobic digestion was observed to have a positive impact on facultative microorganisms, such as Clostridia and Firmicutes, and to improve the hydrolysis stage. Likewise, due to the predominance of aero-tolerant *Methanobacterium* and *Methanosarcina spp.* in the reactor, the methanogenic stage is improved with this pretreatment technique. This technique has been widely experimented with in anaerobic digestion of organic waste to enhance processing stability and efficiency [64]. The studies show that the fundamental mechanism of improving anaerobic digestion through oxygen introduction is the augmentation of the performance of facultative acidogenic and acidogenic microbes under microaerobic conditions [65,66].

The efficiency of microaeration pretreatment on lignocellulose feedstock depends on the dose, exposure time, and frequency [67]. If the feedstock is pretreated before anaerobic digestion, a relatively higher air/oxygen dose is required, but if it is applied during the hydrolysis stage, a low dose is required to avoid methanogenesis obstruction. If desulfurization of the biogas is also targeted, the dose of air/oxygen will be applied according to the percentage of sulfur in the feedstock. Different units of microaeration are available, including O2/LR per day, mg/l, ml/g VS, and μM, which are generally used to define microaeration. However, the general standard technique to define microaeration is still lacking. Due to the low oxygen content added during microaeration, it is challenging to detect with an instrument, as it is utilized by facultative microbes immediately. Therefore, the variation in oxidation-reduction potential (ORP) that differs linearly with the logarithm of oxygen concentration can be used to ascertain the small variation in oxygen concentration in the process [68]. However, the ORP of anaerobic digestion is sensitive to process parameters such as temperature, pH, and ionic strength. Therefore, there is a need for an automatic control and monitoring system that will regulate these parameters. Improper control of oxygen can produce inhibitors that hinder biogas production rather than being advantageous for the methanogenic activities [69]. Data availability on microaerobic pretreatment is crucial to its acceptability and application, especially in developing countries where economic considerations are integral to decision-making.

Microaerobic pretreatment of corn stover demonstrated that the technique facilitated the proliferation of fungi and bacteria, as well as the subsequent hydrolysis and acidogenic stages, thereby enhancing the anaerobic digestion process. Exposure time and pH influenced hydrolysis and methane release, with the optimum methane generation observed at a pH of 8, 12 hrs of exposure time, and 174.3 ml/g TS, resulting in a 16.6% increase in methane released [70]. The effect of microaerobic pretreatment on the biogas yield of buffalo waste was investigated, with a focus on determining the optimum aeration frequency, dose, and exposure time. It was observed that a single injection of 7.3 ml O_2/g VS for 12-hr exposure time degrades 78% additional soluble-acid lignin and enhances biogas yield by 32%. It was reported that when frequency or dose is increased above 7.3 ml O_2/g VS, there was no further increase in lignin degradation [71]. Anaerobic digestion of blackwater was improved with a microaerobic pretreatment technique. Different dosages of oxygen were experimented with, and it was observed that lower dosages (0–10 mg g–1 COD feed) enhanced the hydrolysis rate, whereas higher dosages (50–150 mg g–1 COD feed) led to the accumulation of volatile fatty acids [69]. Microaerobic pretreatment was applied during the anaerobic co-digestion of food waste and brown water, using 5 and 7 ml O_2/day/l. It was observed that pretreatment improved microbial diversity, hydrolysis rate, volatile fatty acid generation, and biogas production [64]. Microaerobic pretreatment is an economical and environmentally friendly pretreatment technique that can help enhance bioenergy generation, particularly in developing countries. However, there is a need to determine whether the microbial diversity of the digester provides more insight into the variations in biochemical processes resulting from microaeration treatment. There is still limited data on the microbiology of anaerobic reactors considered under microaerobic pretreatment.

3.7 BIOAUGMENTATION

Biouaugmentation is a biological pretreatment technique that utilizes microorganisms, such as fungi and bacteria, to break down the complex structure of lignocellulose biomass, thereby enhancing the surface area and accessibility for enzymatic hydrolysis during the biorefinery process [72]. It requires the addition of certain exogenous stress-resistant microbes or a mixture of cultures to the anaerobic digestion microorganism community, enhancing biofuel release. The technique is noted for its ability to improve the hydrolysis and acidogenesis stages during biogas production; however, some investigations have shown that it can also influence the methanogenic stage of the process [73]. The technique breaks down the lignin, opens the cell walls, and hydrolyzes biopolymers. The majority of investigations on bioaugmentation pretreatment are conducted at controlled pH and temperature to support the survival and extended performance of the particular exogenous microbes [74]. Generally, exogenous microorganisms are cellulolytic bacteria that focus on the efficient hydrolysis of complex materials; nonetheless, a few researchers have reported successful inoculation of acidogenic and methanogenic microbes to enhance biogas yield [73–75]. White-rot fungi are noted for their efficient degradation of cellulose and lignin, and bacteria can also have a significant impact [76].

Different microbes, such as *Clostridium cellulolyticum, Caldicellulosiruptor lactoaceticus*, and hemicellulolytic bacteria, have been utilized to degrade cellulose and xylan, respectively, resulting in an increase in biomethane yield of 8–53% [77]. Biomethane yield was increased by 5–70% when different microbial mixtures were applied to brewery spent grains, sweet corn processing wastes, and cowgrass (*Axonopus compressus*) [78–80]. Bioaugmentation of microalgae feedstock for anaerobic digestion with a mixed culture of rumen anaerobic fungi was reported to improve methane yield by 41% [81]. Various strains of anaerobic fungi have been recommended as efficient microbes for commercial-scale applications [82]. Four different cultures of hydrolytic microorganisms were applied to brewery spent grains during anaerobic conditions, and the methane released was improved by 17.8% [83]. Despite the merits of bioaugmentation in feedstock pretreatment, the influence of the technique's culture on the indigenous microorganism community in the digester remains limited, especially due to variations in bioaugmentation activities resulting from different bioaugmentation inocula. Furthermore, the effects of this pretreatment under thermophilic conditions are still limited compared to those under mesophilic conditions due to the higher ammonia stress, the predisposition to environmental alterations, and the unavailability of thermophilic ammonia-resistant consortia.

3.8 FACTORS AFFECTING BIOLOGICAL PRETREATMENT

The merits of biological pretreatment techniques include easy operation, low risk, low energy input, and operating under mild conditions, as well as not producing inhibitory compounds that hinder biogas release [84]. The efficiency of biological pretreatment depends on various factors, including both biotic and abiotic factors. These include biomass type, moisture content, environmental conditions, plant species, particle size, incubation time and temperature, pH, inoculum concentration, aeration rate, etc. [35,84]. Temperature is a crucial factor that impacts the development, physical characteristics, and physical properties of microorganisms [85]. The influence of temperature on microbial development and their enzymatic activities varies significantly among different species. Bacteria can tolerate wider temperature ranges compared to fungi. During microbial activities during pretreatment, microbes release a certain amount of heat as a result of their metabolic activities. Microorganisms are divided into three main groups based on their temperature preference, which includes psychrophiles (0–15 °C), mesophiles (20–45 °C), and thermophiles (60–80 °C) [4]. Mesophilic fungi and bacteria are the most extensively studied due to their ability to grow over a wider range of temperatures. Therefore, the optimum temperature for each microorganism must be established to ensure an efficient process. pH is another crucial factor in the performance of biological pretreatment. It is crucial to the culture medium's growth and metabolic activities by creating an acceptable enzyme-feedstock complex [85]. During microbial incubation, the pH drops after some days of pretreatment, and this change results in the production of lignolytic enzymes, which affects the enzymes' performance [1,86]. It has been observed that both an increase and a decrease in optimum pH significantly reduce the activity of the enzyme. Low pH hinders the performance of cellulases, while higher pH dissolves the enzyme and compromises its performance [2].

Moisture content of the environment is crucial to the microbes' growth and degradation of biomass, and this differs greatly with feedstock and microorganism types [1]. Studies have shown that some fungi and bacteria thrive well when the moisture content is between 40 and 70% on solid feedstocks [87]. In general, single-cell microbes need free water for their propagation and growth; however, higher moisture levels can inhibit aeration and oxygen diffusion, and very low moisture levels can cause a delay in microbial development [88]. The complex microphological arrangements of lignocellulose feedstock have a significant influence on the performance of biological pretreatment. The percentage of phenylpropanoid lignin, cellulose crystallinity, and hemicellulose complexity determines the degradation rate of the microbes. Biological pretreatment is more effective when the lignin is lower compared to higher lignin and with a higher crystallinity index of the feedstock

[89]. For optimal degradation, varying biomass requires different microbial consortia. For biological pretreatment, a longer incubation period is required for effective biomass degradation compared to other conventional pretreatment techniques. The longer the incubation period, the better the ability of the microbes to improve the effectiveness of the process [90]. The particle size of the biomass is also a crucial factor that determines the performance of biological pretreatment. Smaller particles are most suitable for microbial degradation because of the larger surface area for the microbes' activities [2]. Good aeration is required to provide sufficient oxygen for microbes and to assist in carbon dioxide removal, humidity maintenance, and heat dissipation.

3.9 ENVIRONMENTAL AND SOCIAL IMPACTS

Biological pretreatment is one of the most eco-friendly pretreatment techniques, as it utilizes microbes to degrade and hydrolyze lignocellulose biomass [91]. This technique produces more useful by-products compared to non-biological techniques and minimal inhibitory compounds, as it is carried out under mild conditions. Lesser energy and chemical inputs are required, and minimal waste is generated [92]. In addition to the degradation of lignocellulose materials, this method can also be applied to enhance the quality of certain feedstocks by removing unwanted, potentially inhibitory components [93]. Enzymatic pretreatment is mostly preferred over biological pretreatment because microbes and enzymes can be efficiently recycled several times during commercial-scale applications [94]. The use of biological pretreatment to produce enzymes at industrial scales is economical, as commercial enzymes typically require high-tech bioreactors, genetically modified microbes, and specialized purification techniques. The application of microbial consortia in the pretreatment, fermentation, and saccharification of biomass can be utilized in a single step to produce value-added products during a multi-biorefinery process. The amount of residue produced depends on whether the enzymatic or microbial pretreatment is applied; microbial pretreatment releases a lesser quantity of residue than the enzymatic process. However, the residue produced can be used for the production of biomethane and other products [95]. The technique is recognized for its ability to enhance the efficiency of the downstream process, leading to rapid biomass conversion. However, biological pretreatment still has some limitations that are hindering its application, especially at the industrial scale. The high cost of enzymes and microbes is a significant challenge that renders the technique less cost-effective, particularly for industrial applications. Compared to other conventional pretreatment technologies, the process is very slow and requires a longer incubation period, which could result in a waste of productive time. In addition, microbes and enzymes are very sensitive to changes in pH and temperature conditions, which leads to a reduction in the microbes' efficiency or denaturation. Most microbes are highly selective for feedstocks, requiring different microbes for various types of feedstocks, which can complicate the process [96]. To overcome these challenges and utilize the full potential of biological pretreatment, further multidisciplinary research is required to improve existing technologies.

3.10 CONCLUSION

Biological pretreatment is a technique that has been investigated for breaking down the microstructural arrangement of lignocellulose biomass. This technique utilizes microbes to alter the arrangement of lignocellulose biomass, offering an environmentally friendly, energy-efficient, and economical method that can also be applied to the degradation of undesired, toxic materials such as phenolic compounds and furfurals. Biologically pretreated biomass can be saccharified with xylanases and cellulases into sugars that can be converted into biofuels, enzymes, organic acids, and other value-added materials. The performance of biological pretreatment is influenced by both abiotic and biotic factors; therefore, these factors must be considered when selecting microbes for pretreatment. The process requires a higher incubation period, which can be a waste of productive time unless another conventional pretreatment technique is combined. Different optimization approaches have been investigated for the biological pretreatment method, but the effective degree of degradation has not yet been established. Most studies have focused mainly on mesophilic microorganisms; therefore, there is a need to explore thermophilic fungi and bacteria to consolidate biomass bioprocessing. In this case, various microbes can be analyzed and enhanced using genetic and metabolic engineering methods. The secretion of lignocellulolytic enzymes from a single microbe has been attempted through genetic engineering, but a successful outcome has not been achieved in this regard. Consequently, isolating microbes from natural habitats, followed by genetic engineering and co-culturing of microbial consortia in fused bioprocessing, can support the economical generation of value-added products from lignocellulose biomass at the commercial stage. Additionally, the combination of biological pretreatment with other mild physicochemical pretreatment techniques will help meet the targets and strengthen the green economy, providing sustainable and renewable bioproducts globally. Improving the efficiency of biological pretreatment in biorefineries will minimize energy security challenges, lower greenhouse gas emissions, and strengthen the world's bioeconomy.

REFERENCES

[1] S.B. Ummalyma, R.D. Supriya, R. Sindhu, P. Binod, R.B. Nair, A. Pandey, E. Gnansounou, Biological pretreatment of lignocellulosic biomass—Current trends and future perspectives, Second and Third Generation of Feedstocks: The Evolution of Biofuels (2019) 197–212. https://doi.org/10.1016/B978-0-12-815162-4.00007-0.

[2] H.K. Sharma, C. Xu, W. Qin, Biological pretreatment of lignocellulosic biomass for biofuels and bioproducts: An overview, Waste Biomass Valor 10 (2019) 235–251. https://doi.org/10.1007/S12649-017-0059-Y/METRICS.

[3] G. Bedendi, L.D. De Moura Torquato, S. Webb, C. Cadoux, A. Kulkarni, S. Sahin, P. Maroni, R.D. Milton, M. Grattieri, Enzymatic and microbial electrochemistry: Approaches and methods, ACS Meas Sci Au 2 (2022) 517–541. https://doi.org/10.1021/ACSMEASURESCIAU.2C00042/ASSET/IMAGES/LARGE/TG2C00042_0009.JPEG.

[4] J. Vasco-Correa, X. Ge, Y. Li, Biological pretreatment of lignocellulosic biomass, Biomass Fractionation Technologies for a Lignocellulosic Feedstock Based Biorefinery (2016) 561–585. https://doi.org/10.1016/B978-0-12-802323-5.00024-4.

[5] C. Wan, Y. Li, Fungal pretreatment of lignocellulosic biomass, Biotechnol Adv 30 (2012) 1447–1457. https://doi.org/10.1016/J.BIOTECHADV.2012.03.003.
[6] J. Chen, X. Ma, M. Liang, Z. Guo, Y. Cai, C. Zhu, Z. Wang, S. Wang, J. Xu, H. Ying, Physical–chemical–biological pretreatment for biomass degradation and industrial applications: A review, Waste 2 (2024) 451–473. https://doi.org/10.3390/WASTE2040024.
[7] M. Maki, K.T. Leung, W. Qin, The prospects of cellulase-producing bacteria for the bioconversion of lignocellulosic biomass, Int J Biol Sci 5 (2009) 500. https://doi.org/10.7150/IJBS.5.500.
[8] R.R. Singhania, H.A. Ruiz, M.K. Awasthi, C. Di Dong, C.W. Chen, A.K. Patel, Challenges in cellulase bioprocess for biofuel applications, Renew Sustain Energy Rev 151 (2021) 111622. https://doi.org/10.1016/J.RSER.2021.111622.
[9] Y.M. Atiku, S. Abdulsalam, J. Mohammed, S.I. Ahmed, Conversion of cellulosic biomass to bioethanol through fermentation using native microorganisms: A review, J Appl Sci Environ Manage 27 (2023) 1651–1664. https://doi.org/10.4314/JASEM.V27I8.7.
[10] Y.P. Paudel, W. Qin, Characterization of novel cellulase-producing bacteria isolated from rotting wood samples, Appl Biochem Biotechnol 177 (2015) 1186–1198. https://doi.org/10.1007/S12010-015-1806-9/METRICS.
[11] M. Saritha, A. Arora, Lata, Biological pretreatment of lignocellulosic substrates for enhanced delignification and enzymatic digestibility, Indian J Microbiol 52 (2011) 122. https://doi.org/10.1007/S12088-011-0199-X.
[12] R. Chandra, P. Chowdhary, Properties of bacterial laccases and their application in bioremediation of industrial wastes, Environ Sci Process Impacts 17 (2015) 326–342. https://doi.org/10.1039/C4EM00627E.
[13] N. Agarwal, V.S. Solanki, A. Gacem, M.A. Hasan, B. Pare, A. Srivastava, A. Singh, V.K. Yadav, K.K. Yadav, C. Lee, W. Lee, S. Chaiprapat, B.H. Jeon, Bacterial laccases as biocatalysts for the remediation of environmental toxic pollutants: A green and eco-friendly approach—A review, Water 14 (2022) 4068. https://doi.org/10.3390/W14244068.
[14] P. Picart, P.D. De María, A. Schallmey, From gene to biorefinery: Microbial β-etherases as promising biocatalysts for lignin valorization, Front Microbiol 6 (2015) 158855. https://doi.org/10.3389/FMICB.2015.00916/BIBTEX.
[15] U.A. Zitare, F. Vieyra, M.F. Scocozza, F. Rosciani, M.A. Castro, L.O. Martins, D.H. Murgida, Superoxide versus peroxide activation of dye decolorizing peroxidases for bioelectrocatalysis, Bioresour Technol Rep 26 (2024) 101819. https://doi.org/10.1016/J.BITEB.2024.101819.
[16] S. Zhang, T. Xia, J. Wang, Y. Zhao, X. Xie, Z. Wei, X. Zhang, C. Song, X. Song, Role of bacillus inoculation in rice straw composting and bacterial community stability after inoculation: Unite resistance or individual collapse, Bioresour Technol 337 (2021) 125464. https://doi.org/10.1016/J.BIORTECH.2021.125464.
[17] Z. Wu, K. Peng, Y. Zhang, M. Wang, C. Yong, L. Chen, P. Qu, H. Huang, E. Sun, M. Pan, Lignocellulose dissociation with biological pretreatment towards the biochemical platform: A review, Mater Today Bio 16 (2022) 100445. https://doi.org/10.1016/J.MTBIO.2022.100445.
[18] B. Tsegaye, C. Balomajumder, P. Roy, Isolation and characterization of novel lignolytic, cellulolytic, and hemicellulolytic bacteria from wood-feeding termite cryptotermes brevis, Int Microbiol 22 (2019) 29–39. https://doi.org/10.1007/S10123-018-0024-Z.
[19] C. Jiang, Y. Cheng, H. Zang, X. Chen, Y. Wang, Y. Zhang, J. Wang, X. Shen, C. Li, Biodegradation of lignin and the associated degradation pathway by psychrotrophic Arthrobacter sp. C2 from the cold region of China, Cellulose 27 (2020) 1423–1440. https://doi.org/10.1007/S10570-019-02858-3/FIGURES/5.

[20] R. Banu J, S. Sugitha, S. Kavitha, Y. Kannah R, J. Merrylin, G. Kumar, Lignocellulosic biomass pretreatment for enhanced bioenergy recovery: Effect of lignocelluloses recalcitrance and enhancement strategies, Front Energy Res 9 (2021) 679. https://doi.org/10.3389/FENRG.2021.646057/BIBTEX.
[21] S. Bonilla, Z. Choolaei, T. Meyer, E.A. Edwards, A.F. Yakunin, D.G. Allen, Evaluating the effect of enzymatic pretreatment on the anaerobic digestibility of pulp and paper biosludge, Biotechnol Rep 17 (2018) 77–85. https://doi.org/10.1016/J.BTRE.2017.12.009.
[22] Y.X. Liew, Y.J. Chan, S. Manickam, M.F. Chong, S. Chong, T.J. Tiong, J.W. Lim, G.T. Pan, Enzymatic pretreatment to enhance anaerobic bioconversion of high strength wastewater to biogas: A review, Sci Total Environ 713 (2020). https://doi.org/10.1016/J.SCITOTENV.2019.136373.
[23] F. Passos, A. Hom-Diaz, P. Blanquez, T. Vicent, I. Ferrer, Improving biogas production from microalgae by enzymatic pretreatment, Bioresour Technol 199 (2016) 347–351. https://doi.org/10.1016/J.BIORTECH.2015.08.084.
[24] K. Michalska, M. Bizukojć, S. Ledakowicz, Pretreatment of energy crops with sodium hydroxide and cellulolytic enzymes to increase biogas production, Biomass Bioenergy 80 (2015) 213–221. https://doi.org/10.1016/J.BIOMBIOE.2015.05.022.
[25] J.S. Van Dyk, B.I. Pletschke, A review of lignocellulose bioconversion using enzymatic hydrolysis and synergistic cooperation between enzymes—Factors affecting enzymes, conversion and synergy, Biotechnol Adv 30 (2012) 1458–1480. https://doi.org/10.1016/J.BIOTECHADV.2012.03.002.
[26] N. Ben Yahmed, H. Carrere, M.N. Marzouki, I. Smaali, Enhancement of biogas production from Ulva sp. by using solid-state fermentation as biological pretreatment, Algal Res 27 (2017) 206–214. https://doi.org/10.1016/J.ALGAL.2017.09.005.
[27] E. Elbeshbishy, G. Nakhla, H. Hafez, Biochemical methane potential (BMP) of food waste and primary sludge: Influence of inoculum pre-incubation and inoculum source, Bioresour Technol 110 (2012) 18–25. https://doi.org/10.1016/J.BIORTECH.2012.01.025.
[28] R. Ravindran, A.K. Jaiswal, A comprehensive review on pre-treatment strategy for lignocellulosic food industry waste: Challenges and opportunities, Bioresour Technol 199 (2016) 92–102. https://doi.org/10.1016/J.BIORTECH.2015.07.106.
[29] M. Gerhardt, V. Pelenc, M. Bäuml, Application of hydrolytic enzymes in the agricultural biogas production: Results from practical applications in Germany, Biotechnol J 2 (2007) 1481–1484. https://doi.org/10.1002/BIOT.200700220.
[30] L. Yunqin, W. Dehan, W. Lishang, Biological pretreatment enhances biogas production in the anaerobic digestion of pulp and paper sludge, Waste Manag Res 28 (2010) 800–810. https://doi.org/10.1177/0734242X09358734.
[31] S. Yu, G. Zhang, J. Li, Z. Zhao, X. Kang, Effect of endogenous hydrolytic enzymes pretreatment on the anaerobic digestion of sludge, Bioresour Technol 146 (2013) 758–761. https://doi.org/10.1016/J.BIORTECH.2013.07.087.
[32] E. Hosseini Koupaie, S. Dahadha, A.A. Bazyar Lakeh, A. Azizi, E. Elbeshbishy, Enzymatic pretreatment of lignocellulosic biomass for enhanced biomethane production-A review, J Environ Manage 233 (2019) 774–784. https://doi.org/10.1016/J.JENVMAN.2018.09.106.
[33] F.R. Amin, H. Khalid, H. Zhang, S. Rahman, R. Zhang, G. Liu, C. Chen, Pretreatment methods of lignocellulosic biomass for anaerobic digestion, AMB Express 7 (2017) 1–12. https://doi.org/10.1186/S13568-017-0375-4/TABLES/2.
[34] R. Tamilselvan, A. Immanuel Selwynraj, Enhancing biogas generation from lignocellulosic biomass through biological pretreatment: Exploring the role of ruminant microbes and anaerobic fungi, Anaerobe 85 (2024) 102815. https://doi.org/10.1016/J.ANAEROBE.2023.102815.

[35] S.J.A. van Kuijk, A.S.M. Sonnenberg, J.J.P. Baars, W.H. Hendriks, J.W. Cone, Fungal treatment of lignocellulosic biomass: Importance of fungal species, colonization and time on chemical composition and in vitro rumen degradability, Anim Feed Sci Technol 209 (2015) 40–50. https://doi.org/10.1016/J.ANIFEEDSCI.2015.07.026.
[36] K.O. Olatunji, D.M. Madyira, J.O. Amos, Sustainable enhancement of biogas and methane yield of macroalgae biomass using different pretreatment techniques: A mini-review, Energy Environ 35 (2024) 1050–1088. https://doi.org/10.1177/0958305X231193869.
[37] J. Kainthola, A. Podder, M. Fechner, R. Goel, An overview of fungal pretreatment processes for anaerobic digestion: Applications, bottlenecks and future needs, Bioresour Technol 321 (2021) 124397. https://doi.org/10.1016/J.BIORTECH.2020.124397.
[38] A. Civzele, L. Mezule, Isolation and screening of wood-decaying fungi for lignocellulolytic enzyme production and bioremediation processes, Front Fungal Biol 5 (2024) 1494182. https://doi.org/10.3389/FFUNB.2024.1494182/BIBTEX.
[39] W.W. Kululo, N.G. Habtu, M.K. Abera, Z.B. Sendekie, S.W. Fanta, T.A. Yemata, Advances in various pretreatment strategies of lignocellulosic substrates for the production of bioethanol: A comprehensive review, Discov Appl Sci 7 (2025) 1–24. https://doi.org/10.1007/S42452-025-06748-1.
[40] S. Dahiya, A. Rapoport, B. Singh, Biotechnological potential of lignocellulosic biomass as substrates for fungal xylanases and its bioconversion into useful products: A review, Fermentation 10 (2024) 82. https://doi.org/10.3390/FERMENTATION10020082/S1.
[41] Y. Paudel, Two bacillus species isolated from rotting wood samples are good candidates for the production of bioethanol using agave biomass, J Microb Biochem Technol 7 (2015). https://doi.org/10.4172/1948-5948.1000.
[42] F. Zhang, X. Zhao, F. Bai, Improvement of cellulase production in Trichoderma reesei Rut-C30 by overexpression of a novel regulatory gene Trvib-1, Bioresour Technol 247 (2018) 676–683. https://doi.org/10.1016/J.BIORTECH.2017.09.126.
[43] S. Zhou, L.O. Ingram, Synergistic hydrolysis of carboxymethyl cellulose and acid-swollen cellulose by two endoglucanases (CelZ and CelY) from Erwinia chrysanthemi, J Bacteriol 182 (2000) 5676–5682. https://doi.org/10.1128/JB.182.20.5676-5682.2000.
[44] J. Pérez, J. Muñoz-Dorado, T. De La Rubia, J. Martínez, Biodegradation and biological treatments of cellulose, hemicellulose and lignin: An overview, Int Microbiol 5 (2002) 53–63. https://doi.org/10.1007/S10123-002-0062-3.
[45] C. Coelho, L.O. Martins, I. Tiago, Isolation of lignocellulosic biomass-degrading bacteria from Porcellio dilatatus gut-enriched cultures, Appl Microbiol Biotechnol 109 (2025) 1–14. https://doi.org/10.1007/S00253-025-13420-6/FIGURES/4.
[46] R. Millati, S. Syamsiah, C. Niklasson, M. Nur Cahyanto, K. Lundquist, M.J. Taherzadeh, Biological pretreatment: Review, Bioresources 6 (2011) 5224–5259.
[47] A. Zanellati, F. Spina, L. Rollé, G.C. Varese, E. Dinuccio, Fungal pretreatments on non-sterile solid digestate to enhance methane yield and the sustainability of anaerobic digestion, Sustainability 12 (2020) 8549. https://doi.org/10.3390/SU12208549.
[48] J. Vasco-Correa, X. Luo, Y. Li, A. Shah, Comparative study of changes in composition and structure during sequential fungal pretreatment of non-sterile lignocellulosic feedstocks, Ind Crops Prod 133 (2019) 383–394. https://doi.org/10.1016/J.INDCROP.2019.03.043.
[49] E. Rouches, I. Herpoël-Gimbert, J.P. Steyer, H. Carrere, Improvement of anaerobic degradation by white-rot fungi pretreatment of lignocellulosic biomass: A review, Renew Sustain Energy Rev 59 (2016) 179–198. https://doi.org/10.1016/J.RSER.2015.12.317.
[50] C. Zhang, H. Su, J. Baeyens, T. Tan, Reviewing the anaerobic digestion of food waste for biogas production, Renew Sustain Energy Rev 38 (2014) 383–392. https://doi.org/10.1016/J.RSER.2014.05.038.

[51] R. Teixeira Franco, P. Buffière, R. Bayard, Ensiling for biogas production: Critical parameters. A review, Biomass Bioenergy 94 (2016) 94–104. https://doi.org/10.1016/J.BIOMBIOE.2016.08.014.
[52] L. Janke, B.K. McCabe, P. Harris, A. Hill, S. Lee, S. Weinrich, S. Marchuk, C. Baillie, Ensiling fermentation reveals pre-treatment effects for anaerobic digestion of sugarcane biomass: An assessment of ensiling additives on methane potential, Bioresour Technol 279 (2019) 398–403. https://doi.org/10.1016/J.BIORTECH.2019.01.143.
[53] E. Kreuger, I. Nges, L. Björnsson, Ensiling of crops for biogas production: Effects on methane yield and total solids determination, Biotechnol Biofuels 4 (2011) 1–8. https://doi.org/10.1186/1754-6834-4-44/FIGURES/1.
[54] H. Sun, X. Cui, W. Stinner, L. Zhang, X. Ju, J. Guo, R. Dong, Ensiling excessively wilted maize stover with biogas slurry: Effects on storage performance and subsequent biogas potential, Bioresour Technol 305 (2020) 123042. https://doi.org/10.1016/J.BIORTECH.2020.123042.
[55] H. Sun, X. Cui, W. Stinner, G. Mustafa Shah, H. Cheng, S. Shan, J. Guo, R. Dong, Synergetic effect of combined ensiling of freshly harvested and excessively wilted maize stover for efficient biogas production, Bioresour Technol 285 (2019) 121338. https://doi.org/10.1016/J.BIORTECH.2019.121338.
[56] X. Zhao, J. Liu, J. Liu, F. Yang, W. Zhu, X. Yuan, Y. Hu, Z. Cui, X. Wang, Effect of ensiling and silage additives on biogas production and microbial community dynamics during anaerobic digestion of switchgrass, Bioresour Technol 241 (2017) 349–359. https://doi.org/10.1016/J.BIORTECH.2017.03.183.
[57] S.U. Larsen, K. Hjort-Gregersen, A.H. Vazifehkhoran, J.M. Triolo, Co-ensiling of straw with sugar beet leaves increases the methane yield from straw, Bioresour Technol 245 (2017) 106–115. https://doi.org/10.1016/J.BIORTECH.2017.08.117.
[58] J. Guo, X. Cui, H. Sun, Q. Zhao, X. Wen, C. Pang, R. Dong, Effect of glucose and cellulase addition on wet-storage of excessively wilted maize stover and biogas production, Bioresour Technol 259 (2018) 198–206. https://doi.org/10.1016/J.BIORTECH.2018.03.041.
[59] C. Herrmann, C. Idler, M. Heiermann, Biogas crops grown in energy crop rotations: Linking chemical composition and methane production characteristics, Bioresour Technol 206 (2016) 23–35. https://doi.org/10.1016/J.BIORTECH.2016.01.058.
[60] D. Gallegos, H. Wedwitschka, L. Moeller, A. Zehnsdorf, W. Stinner, Effect of particle size reduction and ensiling fermentation on biogas formation and silage quality of wheat straw, Bioresour Technol 245 (2017) 216–224. https://doi.org/10.1016/J.BIORTECH.2017.08.137.
[61] Y. Zhao, J. Yu, J. Liu, H.Y. Yang, L. Gao, X.F. Yuan, Z. jun Cui, X. Wang, Material and microbial changes during corn stalk silage and their effects on methane fermentation, Bioresour Technol 222 (2016) 89–99. https://doi.org/10.1016/J.BIORTECH.2016.09.113.
[62] Q. Chen, W. Wu, D. Qi, Y. Ding, Z. Zhao, Review on microaeration-based anaerobic digestion: State of the art, challenges, and prospectives, Sci Total Environ 710 (2020) 136388. https://doi.org/10.1016/J.SCITOTENV.2019.136388.
[63] A.O. Wagner, N. Lackner, M. Mutschlechner, E.M. Prem, R. Markt, P. Illmer, Biological pretreatment strategies for second-generation lignocellulosic resources to enhance biogas production, Energies 11 (2018) 1797. https://doi.org/10.3390/EN11071797.
[64] J.W. Lim, J.A. Chiam, J.Y. Wang, Microbial community structure reveals how microaeration improves fermentation during anaerobic co-digestion of brown water and food waste, Bioresour Technol 171 (2014) 132–138. https://doi.org/10.1016/J.BIORTECH.2014.08.050.

[65] S. Xu, A. Selvam, J.W.C. Wong, Optimization of micro-aeration intensity in acidogenic reactor of a two-phase anaerobic digester treating food waste, Waste Manag 34 (2014) 363–369. https://doi.org/10.1016/J.WASMAN.2013.10.038.
[66] M. Peces, S. Astals, W.P. Clarke, P.D. Jensen, Semi-aerobic fermentation as a novel pre-treatment to obtain VFA and increase methane yield from primary sludge, Bioresour Technol 200 (2016) 631–638. https://doi.org/10.1016/J.BIORTECH.2015.10.085.
[67] S. Fu, S. Lian, I. Angelidaki, R. Guo, Micro-aeration: An attractive strategy to facilitate anaerobic digestion, Trends Biotechnol 41 (2023) 714–726. https://doi.org/10.1016/J.TIBTECH.2022.09.008/ASSET/3F17E4C3-EFDB-4CA2-AE9E-F8E0EF976EA0/MAIN.ASSETS/GR2.JPG.
[68] D. Nguyen, S.K. Khanal, A little breath of fresh air into an anaerobic system: How microaeration facilitates anaerobic digestion process, Biotechnol Adv 36 (2018) 1971–1983. https://doi.org/10.1016/J.BIOTECHADV.2018.08.007.
[69] B. Guo, N. Yu, D.G. Weissbrodt, Y. Liu, Effects of micro-aeration on microbial niches and antimicrobial resistances in blackwater anaerobic digesters, Water Res 196 (2021) 117035. https://doi.org/10.1016/J.WATRES.2021.117035.
[70] H. Xu, Y. Li, D. Hua, Y. Zhao, L. Chen, L. Zhou, G. Chen, Effect of microaerobic microbial pretreatment on anaerobic digestion of a lignocellulosic substrate under controlled pH conditions, Bioresour Technol 328 (2021) 124852. https://doi.org/10.1016/J.BIORTECH.2021.124852.
[71] I. Zeb, S. Yousaf, M. Ali, A. Yasmeen, A.Z. Khan, J.A. Tariq, Q. Zhao, A.M. Abbasi, R. Ahmad, T.M. Khalil, A. Yaqoob, M. Bilal, In-situ microaeration of anaerobic digester treating buffalo manure for enhanced biogas yield, Renew Energy 181 (2022) 843–850. https://doi.org/10.1016/J.RENENE.2021.09.089.
[72] K.O. Donkor, L.D. Gottumukkala, R. Lin, J.D. Murphy, A perspective on the combination of alkali pre-treatment with bioaugmentation to improve biogas production from lignocellulose biomass, Bioresour Technol 351 (2022) 126950. https://doi.org/10.1016/J.BIORTECH.2022.126950.
[73] Ç. Akyol, O. Ince, M. Bozan, E.G. Ozbayram, B. Ince, Fungal bioaugmentation of anaerobic digesters fed with lignocellulosic biomass: What to expect from anaerobic fungus Orpinomyces sp., Bioresour Technol 277 (2019) 1–10. https://doi.org/10.1016/J.BIORTECH.2019.01.024.
[74] A. Nzila, Mini review: Update on bioaugmentation in anaerobic processes for biogas production, Anaerobe 46 (2017) 3–12. https://doi.org/10.1016/J.ANAEROBE.2016.11.007.
[75] J. Zhang, R.B. Guo, Y.L. Qiu, J.T. Qiao, X.Z. Yuan, X.S. Shi, C.S. Wang, Bioaugmentation with an acetate-type fermentation bacterium Acetobacteroides hydrogenigenes improves methane production from corn straw, Bioresour Technol 179 (2015) 306–313. https://doi.org/10.1016/J.BIORTECH.2014.12.022.
[76] K.O. Donkor, L.D. Gottumukkala, R. Lin, J.D. Murphy, A perspective on the combination of alkali pre-treatment with bioaugmentation to improve biogas production from lignocellulose biomass, Bioresour Technol 351 (2022) 126950. https://doi.org/10.1016/J.BIORTECH.2022.126950.
[77] X. Peng, R.A. Börner, I.A. Nges, J. Liu, Impact of bioaugmentation on biochemical methane potential for wheat straw with addition of Clostridium cellulolyticum, Bioresour Technol 152 (2014) 567–571. https://doi.org/10.1016/J.BIORTECH.2013.11.067.
[78] J.T.E. Lee, Q. Wang, E.Y. Lim, Z. Liu, J. He, Y.W. Tong, Optimization of bioaugmentation of the anaerobic digestion of Axonopus compressus cowgrass for the production of biomethane, J Clean Prod 258 (2020) 120932. https://doi.org/10.1016/J.JCLEPRO.2020.120932.

[79] A. Martin-Ryals, L. Schideman, P. Li, H. Wilkinson, R. Wagner, Improving anaerobic digestion of a cellulosic waste via routine bioaugmentation with cellulolytic microorganisms, Bioresour Technol 189 (2015) 62–70. https://doi.org/10.1016/J.BIORTECH.2015.03.069.

[80] M. Čater, L. Fanedl, Š. Malovrh, R. Marinšek Logar, Biogas production from brewery spent grain enhanced by bioaugmentation with hydrolytic anaerobic bacteria, Bioresour Technol 186 (2015) 261–269. https://doi.org/10.1016/J.BIORTECH.2015.03.029.

[81] S. Aydin, E. Yıldırım, O. Ince, B. Ince, Rumen anaerobic fungi create new opportunities for enhanced methane production from microalgae biomass, Algal Res 23 (2017) 150–160. https://doi.org/10.1016/J.ALGAL.2016.12.016.

[82] S.S. Dagar, S. Kumar, P. Mudgil, A.K. Puniya, Comparative evaluation of lignocellulolytic activities of filamentous cultures of monocentric and polycentric anaerobic fungi, Anaerobe 50 (2018) 76–79. https://doi.org/10.1016/J.ANAEROBE.2018.02.004.

[83] M. Čater, L. Fanedl, Š. Malovrh, R. Marinšek Logar, Biogas production from brewery spent grain enhanced by bioaugmentation with hydrolytic anaerobic bacteria, Bioresour Technol 186 (2015) 261–269. https://doi.org/10.1016/j.biortech.2015.03.029.

[84] S.K. Singh, Biological treatment of plant biomass and factors affecting bioactivity, J Clean Prod 279 (2021) 123546. https://doi.org/10.1016/J.JCLEPRO.2020.123546.

[85] Anu, V. Kumar, D. Singh, B. Singh, A greener, mild, and efficient bioprocess for the pretreatment and saccharification of rice straw, Biomass Convers Biorefin 13 (2023) 4121–4133. https://doi.org/10.1007/S13399-021-01450-9/FIGURES/7.

[86] L.M. Marra, S.M. de Oliveira-Longatti, C.R.F.S. Soares, J.M. de Lima, F.L. Olivares, F.M.S. Moreira, Initial pH of medium affects organic acids production but do not affect phosphate solubilization, Braz J Microbiol 46 (2015) 367–375. https://doi.org/10.1590/S1517-838246246220131102.

[87] M. Raimbault, General and microbiological aspects of solid substrate fermentation, EJB Electron J Biotechnol 1 (1998) 717–3458. http://www.mpl.orstom.fr (accessed June 4, 2025).

[88] K.S.M.S. Raghavarao, T.V. Ranganathan, N.G. Karanth, Some engineering aspects of solid-state fermentation, Biochem Eng J 13 (2003) 127–135. https://doi.org/10.1016/S1369-703X(02)00125-0.

[89] S. Meenakshisundaram, A. Fayeulle, E. Léonard, C. Ceballos, X. Liu, A. Pauss, Combined biological and chemical/physicochemical pretreatment methods of lignocellulosic biomass for bioethanol and biomethane energy production—A review, Appl Microbiol 2 (2022) 716–734. https://doi.org/10.3390/APPLMICROBIOL2040055.

[90] U. Brémond, R. de Buyer, J.P. Steyer, N. Bernet, H. Carrere, Biological pretreatments of biomass for improving biogas production: An overview from lab scale to full-scale, Renew Sustain Energy Rev 90 (2018) 583–604. https://doi.org/10.1016/j.rser.2018.03.103.

[91] C.H. Chen, J.Y. Yao, B. Yang, H.L. Lee, S.F. Yuan, H.Y. Hsieh, P.H. Liang, Engineer multi-functional cellulase/xylanase/β-glucosidase with improved efficacy to degrade rice straw, Bioresour Technol Rep 5 (2019) 170–177. https://doi.org/10.1016/J.BITEB.2019.01.008.

[92] W. Zhong, Z. Zhang, Y. Luo, S. Sun, W. Qiao, M. Xiao, Effect of biological pretreatments in enhancing corn straw biogas production, Bioresour Technol 102 (2011) 11177–11182. https://doi.org/10.1016/J.BIORTECH.2011.09.077.

[93] O. Parthiba Karthikeyan, E. Trably, S. Mehariya, N. Bernet, J.W.C. Wong, H. Carrere, Pretreatment of food waste for methane and hydrogen recovery: A review, Bioresour Technol 249 (2018) 1025–1039. https://doi.org/10.1016/J.BIORTECH.2017.09.105.

[94] Anu, A. Kumar, A. Rapoport, G. Kunze, S. Kumar, D. Singh, B. Singh, Multifarious pretreatment strategies for the lignocellulosic substrates for the generation of renewable

and sustainable biofuels: A review, Renew Energy 160 (2020) 1228–1252. https://doi.org/10.1016/J.RENENE.2020.07.031.

[95] R.C. Kuhad, A. Rapoport, V. Kumar, D. Singh, V. Kumar, S.K. Tiwari, S. Ahlawat, B. Singh, Biological pretreatment of lignocellulosic biomass: An environment-benign and sustainable approach for conversion of solid waste into value-added products, Crit Rev Environ Sci Technol 54 (2024) 771–796. https://doi.org/10.1080/10643389.2023.2277670.

[96] R.S. Abolore, S. Jaiswal, A.K. Jaiswal, Green and sustainable pretreatment methods for cellulose extraction from lignocellulosic biomass and its applications: A review, Carbohydr Polym Technol Appl 7 (2024) 100396. https://doi.org/10.1016/J.CARPTA.2023.100396.

4 Chemical pretreatment

4.1 INTRODUCTION

Chemical pretreatment is the process of breaking down the crystalline arrangement of cellulose and eliminating the lignin and hemicellulose from lignocellulose biomass by applying different chemical agents such as acid, alkali, oxidizing agents, ionic liquids, etc. [1,2]. This technique can remove, redistribute, or modify condensed and degraded matter through the action of the applied chemical. Lignin alteration, which includes interunit bond cleavage, functional group modification, and breakdown of lignin monomers, is presented in Figure 4.1. These are the primary changes noticed from the chemical pretreatment technique. Some of the modifications in lignin arrangement are interrelated; for instance, ether bond cleavage is usually achieved with the generation of extra phenolic OH and removal of aliphatic OH groups. It has been observed that these morphological alterations will influence the adsorption between enzymes and lignin, thereby enhancing the subsequent enzymatic hydrolysis of pretreated feedstock [3]. This technique is one of the most promising methods to enhance the biodegradation of cellulose by eliminating lignin and/or hemicellulose, thereby reducing the polymerization degree and cellulosic crystallinity of the biomass [4]. To achieve efficient chemical pretreatment, enough catalyst and/or pretreatment chemical loadings are crucial; therefore, the use of a higher solvent-to-biomass ratio is always required. The efficiency of chemical pretreatment depends on the chemical used, process parameters (including exposure time, temperature, solvent concentration, and feedstock-to-solvent ratio), and the operation mode (batch, semi-batch, continuous, etc.) [5]. Chemical pretreatment has been reported to offer better operational efficiency, making it a popular technique at the industrial scale [6]. The selection of the most suitable chemical techniques depends primarily on the type of biomass, the intended final product, economic considerations, and environmental impact, as some chemicals produce inhibitory materials that are harmful to the environment. This chapter provides an overview of the various types of chemical pretreatments available and their impact on lignocellulosic biomass. Modification in the microstructural arrangement and conversion efficiency of each technique was discussed. The economy, merits, and limitations of these techniques will be highlighted, with future studies in the area recommended.

4.2 ACIDIC

The acidic pretreatment technique involves using concentrated or dilute solvents to break down the strong arrangement of lignocellulose biomass. During this process, the acid serves as a catalyst that modifies the intramolecular and intermolecular linkages within lignin and hemicellulose and hydrolyzes the carbohydrates in the biomass, especially hemicellulose [8]. The target of this technique is to improve the feedstock's sugar digestibility by enhancing hemicellulose solubility, promoting

DOI: 10.1201/9781003633815-7

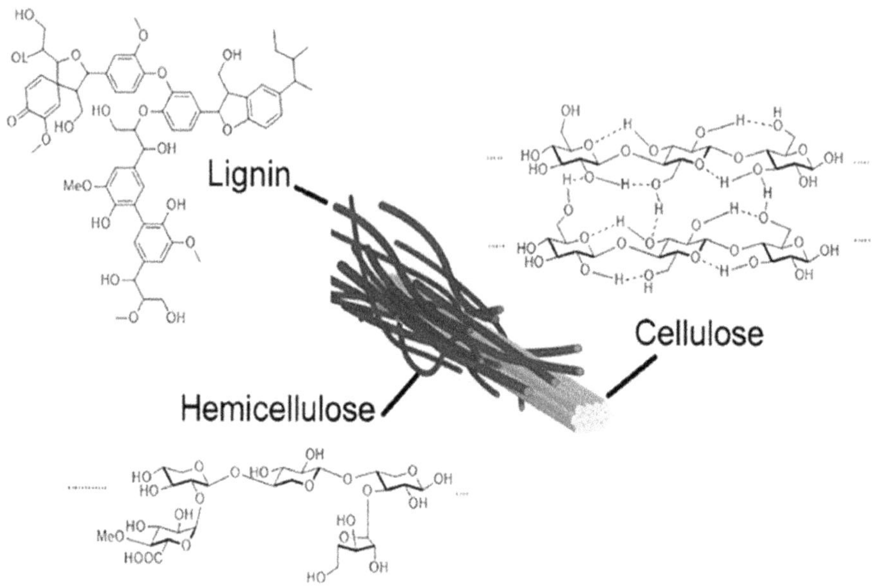

FIGURE 4.1 Effects of chemical pretreatment on lignocellulose biomass [7].

lignin precipitation, and exposing cellulose to facilitate microorganisms' accessibility. The bonds between the long-chain hemicellulose and cellulose are broken down through the hydronium ion produced during acid catalysis [9]. It is a widely studied pretreatment technique that is closer to commercial application. Several inorganic acids, such as hydrochloric acid, sulfuric acid, phosphoric acid, nitric acid, etc., and organic acids such as oxalic acid, glycolic acid, citric acid, maleic acid, acetic acid, levulinic acid, and formic acid have been experimented with for the lignocellulose biomass pretreatment [10–12]. The technology can be carried out either with diluted acid at higher temperatures (>200 °C) or concentrated acid at mild temperatures (< 50 °C) [13]. These pretreatment approaches have their merits and challenges. For instance, a lower acid concentration is required for dilute-acid pretreatment; however, the process requires more energy due to the higher temperatures needed. However, concentrated acid pretreatment requires lower energy since a lower temperature is needed, but a higher volume of acid is required, and the tendency to release inhibitory compounds (phenolic, furfural, and 5-hydroxymethylfurfural) is very high. These inhibitory compounds can have a negative influence on the downstream bioprocess yield [14]. Therefore, the selection of acid type, concentration, exposure time, solid-to-liquid loading ratios, and temperature is crucial to the successful application of acidic pretreatment.

Acid pretreatment is known for its ability to disrupt the lignocellulose matrix via the cleavage of glucosidic links, which results in the conversion of polysaccharides into monomeric and oligomeric sugars [15]. Hemicellulose is the most affected component of lignocellulose feedstock during acid pretreatment, while certain percentages of cellulose and lignin are also solubilized to release carbohydrates and oligomers [16]. The lignin content undergoes primarily acid-catalyzed depolymerization and condensation, producing small soluble portions and insoluble residue solids. These chemical modifications in lignin can have a significant effect on the downstream bioprocessing of pretreated biomass. The hydrolysate soluble contents primarily consist of oligomeric, monomeric, and dimeric phenolic compounds released from lignin [17,18]. The products of hydrolysate soluble degeneration can have inhibitory impacts on bioprocessing; therefore, they need further treatment.

Inorganic acids such as hydrochloric acid and sulfuric acid have been the most widely studied pretreatment techniques; however, recent studies have reported the efficiency of organic acids as solvents for biomass pretreatment. Organic acid pretreatment is a new and effective clean separation technique, offering the merits of higher acid separation efficiency, mild reaction conditions, and reduced pollution [19]. Because of the low coefficient of organic acid acidity, hemicellulose content is separated selectively, and the lignin can be degraded effectively due to the solubility resemblance between lignin and some of the organic acids. It has been observed that acidic pretreatment using formic and acetic acid has a significant effect on lignin removal [20,21]. Under mild conditions, organic acids are selective and have a minimal effect on cellulose and glucose degradation, which makes them efficient for releasing hemicellulosic sugars with lower product degradation [22]. Compared to inorganic acids, organic acids are less toxic, do not produce an offensive odor, and acid recovery is easier, which is a major advantage in sustainable biorefineries [23]. Several studies have applied different acidic pretreatments on varying lignocellulose biomass to enhance bioenergy production, as shown in Table 4.1. It can be noticed from the table that acidic pretreatment has a significant influence on bioenergy products. However, it was observed that the influence of acid pretreatment varied based on the microstructural arrangement of the feedstock. Therefore, there is a need to establish a unified standard that matches the acidic pretreatment conditions to lignocellulose biomass based on their morphological arrangement. This will improve the rate of acceptability at the industrial level, especially when the plant does not depend on a single biomass for its production. Despite the strength of acidic pretreatment in degrading the lignocellulose biomass matrix and improving downstream bioprocessing, there are some limitations that need to be addressed to make the process more economically viable. The release of inhibitory compounds such as ketones, phenolic acids, and aldehydes during lignin and sugar degradation is regarded as a major limitation that needs to be addressed urgently [24]. The need for equipment with higher corrosion resistance, due to the corrosive properties of the acid, is another area that requires attention.

4.3 ALKALI

The alkali pretreatment technique is a process that involves applying a base to lignocellulose biomass to reduce the recalcitrant properties of the feedstock and make cellulose more accessible during the downstream process [35]. Sodium hydroxide, calcium hydroxide, sodium carbonate, potassium hydroxide, hydrogen peroxide, and aqueous ammonia are some of the common alkali solvents used in biomass

TABLE 4.1
Influence of different acidic pretreatments on lignocellulose biomass

Feedstock	Treatment conditions	Biorefinery process	Observation/impacts	Reference
Olive tree biomass	2% w/v H_2SO_4 at 130 °C for 90 min.	Enzymatic hydrolysis at 50 °C and 5.0 pH for 72 hrs.	The sugar yield was enhanced by 34%.	[25]
Rice husk	5% v/v H_2SO_4 at 121 °C for 30, 60, and 90 min.	Dark fermentation using serum bottles at 35 °C and a pH of 7–7.5.	The optimum sugar yield was obtained at 60 min, but it does not translate to hydrogen release due to a higher concentration of 5-hydroxymethylfurfural.	[26]
Triarrhena lutarioriparia	10% w/v HCL at 120 °C for 1 hr.	Enzymatic hydrolysis at 50 °C and 4.8 pH for 72 hrs.	The reducing sugar was improved by 34% compared with the untreated substrate.	[27]
Arachis hypogea shells	0.5% v/v H_2SO_4 at 90 °C for 15 min.	Batch anaerobic digestion at 37 °C for 35 days.	Biogas yield was increased by 13%, and biomethane increased by 178%.	[28]
Wheat straw	1% H_2SO_4 + 2.4% Na_2SO_3 at 180 °C for 30 min.	Enzymatic hydrolysis at 55 °C for 48 hrs.	Ethanol production increased 350% compared to the untreated feedstock. The addition of Na_2SO_3 was able to reduce the inhibitory compound, but above 2.4%, the yield started to decline.	[29]
Rice straw	5.01% w/w citric acid at 135.91 °C for 30.86 min.	Batch anaerobic digestion at 35 °C for 45 days.	The biogas yield increased by 639.58%.	[30]
Sorghum bicolor stalk	0.75% v/v H_2SO_4 at 118 °C for 52 min.	Batch anaerobic digestion at 37 °C for 28 days.	The methane yield was reduced from 55 to 54.5%.	[31]
Sugarcane bagasse	0.2% w/v H_3PO_4 at 186 °C for 24 min.	Enzymatic hydrolysis at 50 °C and 4.8 pH for 72 hrs.	Sugar released increased by 98%.	[32]
Rapeseed straw	1% H_2SO_4 at 180 °C for 10 min.	Fermentation using Pyrex flasks at 30 °C and 5.4 pH for 48 hrs.	90% ethanol yield was recorded from the pretreated substrate, while only 32% was obtained from the untreated substrate.	[33]
Municipal solid waste	HCl with a pH of 6–1 for 24 hrs.	Batch anaerobic digestion at 37 °C for 30 days.	The cumulative biogas released was enhanced by 13.2–28.9%, with the maximum improvement observed at pH 3.	[34]
Municipal solid waste	CH_3COOH with a pH of 6–1 for 24 hrs.	Batch anaerobic digestion at 37 °C for 30 days.	The total biogas yield improved by 8.2–16%, with an optimum yield at a pH of 1.	[34]

pretreatment [36,37]. The process can be carried out at lower temperatures and pressures, but the exposure time can be longer, ranging from hours to days or weeks, depending on the microstructural arrangement of the biomass [38]. Alkaline pretreatment can be carried out by dipping the lignocellulose biomass in an alkali solution for a certain period at a specified temperature. The first reactions during alkali pretreatment are saponification and solvation, which result in biomass swelling and increase the surface area, thereby improving accessibility for microorganisms [39]. The technique primarily modifies ester and glycosidic side chains, resulting in partial deconstruction of cellulose, hemicellulose solvates, and alteration of lignin arrangement [40]. It dissociates or separates the biomass into subsequent lignocellulosic fibers. When a strong alkali solution is used, it causes dissolution, peeling of end groups, hydrolysis, degradation, and decomposition of dissolved polysaccharides [40]. During alkaline hydrolysis, the saponification of intermolecular ester bonds, as well as the cross-linking of xylan hemicellulose and other constituents, is broken, resulting in a porous structure due to the degradation of the cross-linkages. Alkali application to lignocellulose biomass eliminates the acetyl groups of hemicelluloses, reduces the steric hindrance of hydrolytic enzymes, and enhances the degradation of carbohydrates. Lignin removal from the cell wall is primarily related to hydrolysis [12]. The lignin arrangement is disrupted, and the bonds (aryl-ether, ester, and C=C bonds) related to lignin and carbohydrate polymers are altered during alkali pretreatment [41].

The base-catalyzed cleavage of ether bonds (β- and α-ethers) is the major reaction of lignin during alkaline pretreatment. α-ethers are comparatively stable during alkali pretreatment, while phenolic α-ethers are affected by pretreatment by cleaving to form a quinone methide intermediate. The non-phenolic α-ethers can be cleaved if a free hydroxyl group is available close to the β position, where the adjacent OH groups can ease the cleavage through the generation of epoxide [42]. The quinone methide is an unstable compound that later reacts to produce a more stable material. Furthermore, the hydroxide ion can degrade γ-hydroxyl groups, resulting in the loss of γ-carbon as formaldehyde via an inverse aldol condensation process and generating a vinyl ether. Lignin condensation during alkali pretreatment can also lead to the production of new alkaline-stable C–C links. The α-carbon of quinone methide released from the generation of the OH from the α-point of phenolic arrangement can be nucleophilically attacked by the C1 or C5 carbon of other phenoxide subarrangement to produce α-1 or α-5 class of condensation material, respectively [43]. Compared to acidic pretreatment, alkali is less corrosive, making it applicable with less expensive materials that have lower corrosion resistance. Chemical recovery and reuse are possible in alkaline pretreatment [44]. Sodium and calcium hydroxide pretreatments are the most common alkali pretreatments, which have been observed for their strength to degrade inter- and intra-unit links and break the lignin significantly and a certain amount of hemicelluloses [45,46]. The use of potassium permanganate solution (APP) as an alkali pretreatment solvent was observed to degrade feedstock through oxidation with higher carbohydrate contents [47]. Recently, the investigations on alkali pretreatment have focused on process optimization of concentration, time, and temperature optimization, morphological analysis of pretreatment impacts, and the influence on subsequent downstream processes [48,49]. Temperature, exposure time, and alkali concentration are the primary process parameters that influence the performance of this pretreatment. Using excessive concentrations or volumes of alkali can lead to the generation of inhibitory compounds and have a harmful impact on the environment [50]. The economy of the technique can be improved by solvent recovery after pretreatment, which will also reduce the harmful environmental impacts. One of the major limitations of this technique is the longer exposure time, which can be addressed by increasing the process temperature or by combining it with other pretreatment techniques, such as thermal pretreatment or particle size reduction [51]. Another limitation of this technique is its lower efficiency on biomass with a higher lignin content, such as wood. However, the outcome of alkali pretreatment needs to be weighed against its impacts on the ease of operation, the economy of its biorefinery, and the trade-off between different expenses, such as feedstock cost, capital investment, and operational costs [52]. Alkali pretreatment has been successfully applied to various lignocellulose biomass, as shown in Table 4.2. It can be observed that the optimum pretreatment conditions for different biomasses vary, and these depend on the microstructural arrangement of the feedstock. Some of the alkali pretreatment technologies are still in the laboratory stage, and there is a need to examine them at the prototype and commercial scales to make the findings available and encourage adoption and further investment in the bioenergy sector.

4.4 OXIDATIVE

The oxidative pretreatment technique involves the application of oxidants to enhance the degradation of recalcitrant characteristics, thereby making hemicellulose and cellulose more accessible during downstream bioprocessing. The process leads to lignin cleavage and hemicellulose degradation into organic acids and constituent sugars, with partial breakdown of cellulose [61]. Oxygen, hydrogen peroxide, iron chloride, and ozone are the primary oxidizing agents applied for lignocellulose delignification [62]. These oxidants can be dissolved in water at varying concentrations and applied to the lignocellulose biomass, with the influence depending on the oxidant type, exposure time, temperature, and biomass type [63]. Different reactions take place during oxidative pretreatment. This includes side-chain degradation, electrophilic replacement, and cleavage of alkyl–aryl ether bonds, as well as oxidative cleavage of aromatic nuclei. The degree of degradation is defined by the level of reactivity of oxidizing agents with the aromatic rings of biomass lignin. This technique oxidizes lignin polymers into carboxylic acids that can become inhibitory compounds during the downstream bioprocessing stage [64]. The hemicellulose contents of the biomass are also influenced by this technique. The crystalline structure of cellulose biomass can also be opened during oxidative pretreatment.

When oxygen is used, it improves the reaction process and production of radicals at high pressure and temperature. The reaction of the hydroxyl group of lignin with alkali and oxygen releases the phenolate ion, an intermediate referred to as hydroperoxide. This intermediate is later degraded to produce lignin contents that are soluble in water. Utilization of pure oxygen can enhance the reaction rate; however, the operational cost is high [13]. The application of H_2O_2 results in lignin solubilization and decomposes lignin from biomass, thereby increasing the surface area for subsequent

TABLE 4.2
Performance of alkali pretreatment on different biomasses

Feedstock	Treatment conditions	Biorefinery process	Observation /impacts	Reference
Wheat straw	1.6% w/w NaOH at 30 °C for 24 min.	Biomethane potential (BMP) testing using serum bottles for 40 days at 37 ± 2 °C.	Lignin and xylan were reduced by 36 and 37%, respectively, while glucan was increased by 16%. Biomethane yield was improved by 104%.	[53]
Wheat straw	6% w/w CaO at 121 °C for 60 min.	Batch hydrogen fermentation at 37 °C for 10 days.	There is no significant influence on cellulose; hemicellulose is reduced by 14%, lignin is reduced by 12%, and hydrogen is increased by 164%.	[54]
Rice straw	7% w/v NaOH at –12 °C for 3 min.	Batch fermentation in 100 ml serum bottles at 105 °C at pH 7.	Cellulose increased by 19%, hemicellulose increased by 16%, lignin decreased by 42.9%, and biohydrogen increased by 161.92%.	[55]
Arachis hypogea shells	3% w/w NaOH at 90 °C for 15 min.	Batch test using automatic methane potential testing system (AMPTS) II for 30 days at 37 ± 2 °C.	Lignin, hemicellulose, and cellulose broke down, and the biomethane yield increased by 70%.	[56]
Goose weed	1% w/w NaOH at 50 °C for 24 hrs.	Fermentation at 33 °C using 2% w/v immobilized yeast for 9 days.	The cellulase enzyme breaks and releases reducing sugar of up to 50% of the total sugar.	[57]
Wheat straw	6% w/w CaO_2 at 121 °C for 60 min.	Batch hydrogen fermentation at 37 °C for 10 days.	The cellulose content increased by 2.8%, while hemicellulose and lignin were reduced by 9 and 18%, respectively. The production of biohydrogen increased by 66%.	[54]
Sugarcane bagasse	0.7% w/v NaOH at 70 °C for 3 hrs.	Enzymatic hydrolysis at 50 °C for 72 hrs and 4.8 pH.	Delignification and glucan recovery were 44.67 and 97.14%, and ethanol production increased by 13%.	[58]
Rice straw	3% m/v KOH at 70 °C for 3 hrs.	Fermentation with Saccharomyces cerevisiae strain at 30 °C for 48 hrs and a pH of 4.8.	The sugar yield was enhanced by 92%, while bioethanol release improved by 87%.	[59]
Wheat straw	0.2 mol/l NaOH at 30 °C for 5 hrs.	Enzymatic hydrolysis at 50 °C for 72 hrs and a pH of 5.	The xylan and glucan conversion rates were recorded at 49.4 and 44.6%, respectively.	[60]
Xyris capensis	4% w/w NaOH at 90 °C for 20 min.	Batch test using AMPTS II for 35 days at 37 ± 2 °C.	Scanning electron microscope (SEM) analysis reveals a severe damage image of lignin, with a corresponding increase in methane yield of 143%.	[49]

improvement in downstream bioprocesses [65]. Lignin oxidation into aromatic compounds can behave as inhibitory materials that will negatively affect the subsequent stage of biorefinery. Reports have shown that H_2O_2 concentrations exceeding 4% can result in a higher release of hydroxyl ions and become an inhibitory compound during anaerobic digestion [66]. The addition of Na_2CO_3 during oxidation pretreatment can help reduce the potential for inhibitor release [67]. Application of oxidative pretreatment on Sorghum bicolor stalk was observed to reduce lignin and hemicellulose contents by 73 and 42%, respectively, and cellulose accessibility was enhanced by 23%. When the pretreated substrate was subjected to anaerobic digestion, the biomethane yield was improved by 56 and 65% compared to the untreated and H_2SO_4-pretreated samples, respectively [31]. Oxidative pretreatment of agricultural residues using Fenton (Fe II) and ozone (O_3) was noticed to enhance biomethane release and digestion efficiency by around 25% [68]. Oxidative pretreatment of *Xyris capensis* using 75% H_2O_2 and 25% H_2SO_4 was investigated, and the SEM image shows a significant influence of the pretreatment on the morphological arrangement of the feedstock, and biomethane released was increased by 48.52% [69]. Oil palm empty fruit bunches were pretreated with oxidative pretreatment using 6% oxygen loading at 180 °C for 45 min and digested under mesophilic and thermophilic conditions. It was observed that the biomethane yield increased by 43 and 32% under mesophilic and thermophilic conditions, respectively. However, the rate of release under thermophilic conditions is higher than under mesophilic conditions [70]. Oxidative pretreatment of birch wood using Fenton reagent (0.001 M $FeCl_3$ + 0.01 M H_2O_2) was reported to reduce the biogas yield by 18% [71]. This technique is challenging at the industrial scale, and a higher concentration of alkali or acid is required. The strength of the oxidizing agent to degrade a significant amount of hemicellulose and reduce the available organic matter if the process parameters are not properly selected is another drawback of this technique.

4.5 OZONOLYSIS

Ozonolysis pretreatment is the application of ozone to break down the lignocellulose materials with minimal influence on hemicellulose and cellulose. Ozone is a strong oxidizing agent (E^0 = 2.07 V at 25 °C) that is soluble in water (110 mg/l at 25 °C) and available for use after it is released from oxygen in an endothermic reaction [72]. It is a highly reactive, non-linear triatomic molecule toward compounds with double bonds and functional groups that have high electron densities. The process degrades primarily from lignin, which is an electron-rich component, rather than carbohydrate, because of the electron shortage in the terminal oxygen of ozone. Lignin also has a higher content of double bonds, which can make it the most oxidized component during pretreatment [73]. Ozone has been widely used as a bleaching agent in the pulp and paper industry, and it has recently been investigated as a biomass pretreatment method. Ozone pretreatment has no significant influence on downstream bioprocessing, such as fermentation and enzymatic hydrolysis, and generates insignificant inhibitory compounds [74]. The reaction of C=C bond segmentation follows the Criegee mechanism, which predicts the ozonolysis of alkene compounds (C=C) in three distinct stages. The process is fast, as demonstrated by rapid lignin debasement and high initial ozone consumption rates [73]. In this technique, soluble compounds with low molecular weights, primarily organic acids such as acetic and formic acids, are generated, resulting in a reduction in the solution's acidification. Generally, ozone pretreatment consists primarily of an ozone generator, a reactor where the biomass and ozone react, and a tail gas treatment unit where the unused ozone is quenched. This type of setup is an in situ equipment that can be manufactured or assembled on-site when needed, reducing or eliminating handling and transport costs, as well as other storage-related expenses [75]. This method is gaining more attention due to the low level of inhibitory compounds released and the significant improvement in cellulose enzymatic hydrolysis rate [76].

The performance of ozonolysis pretreatment is determined by various parameters, including ozone-to-air flow rate, reaction solvent, pH, reactor type, catalyst used, reaction time, and biomass characteristics such as particle size and moisture content [77]. Biomass type and composition have no significant influence on ozone absorption, as the process is primarily controlled by the diffusion of ozone through the pores of the biomass [78]. The moisture content of the substrate is a major factor influencing the impact of ozonolysis pretreatment, as moisture is required to enhance ozone diffusion from the gas phase to the feedstock surface, facilitating efficient reaction with lignin. Lower moisture content limits the mass transfer of ozone because it cannot be solubilized in water and diffuse into the biomass surface. Likewise, if the moisture content is too high, it can form a thick water film that blocks the pores on the feedstock surface and increases the resistance to mass transfer [79]. At relatively high moisture content, a longer exposure time to ozone can be attained; however, this can encourage the release of non-selective hydroxyl radicals, which can result in the loss of carbohydrates [80]. Biomass particle size is also a crucial parameter that influences the performance of ozonolysis pretreatment, as the reaction occurs on the surface of the feedstock. It was observed that the degradation effect is more significant when relatively smaller particles are used. But the degradation of carbohydrate contents is higher when smaller biomass particles are considered. Ozonolysis pretreatment has been investigated on various biomasses, including sugarcane bagasse [79], maize stover [80], wheat straw [81], and agroindustry [82]. It was observed that it has the capacity to improve the downstream bioprocessing of lignocellulose biomass. However, different optimum process parameters have been reported based on the varying feedstocks considered. It is therefore necessary to investigate the process using a more sophisticated optimization technique and perform a kinetic analysis of the ozonolysis pretreatment. The industrial application of this pretreatment technique for various lignocellulose biomass is required, as most of the present study is still at the laboratory scale.

4.6 SPORL

SPORL pretreatment is a lignocellulose biomass pretreatment that utilizes the fundamental knowledge of sulfite pulping, which involves combining sulfite-based salts to treat biomass at high temperatures before size reduction [83]. Sulphite salts of Na, Ca, or Mg origin can be used to pretreat lignocellulose biomass at a pH of around 2–4, with temperatures of 160–180 °C for a short period (e.g., 30 min) [84]. Particle size reduction (e.g., disk milling) is the subsequent operation after sulfite treatment to produce a fibrous feedstock for the subsequent fermentation, digestion, or saccharification. Sulphite salts are industrial chemicals with a strong biocidal influence and can degrade hemicelluloses; they can also be produced from waste containing sulfite. It is a bright pretreatment technique for the conversion of woody biomass, softwoods, grass, and microalgae [85]. Sulphite pretreatment has been experimented with successfully in sewage sludges for some time [86]. However, most lignocellulose biomass has a stronger cell wall with higher hemicellulose and cellulose contents, which will make the influence of SPORL pretreatment differ among them. SPORL pretreatment was applied to spruce wood, and the cumulative yield of 87.9% xylose and glucose was recorded after saccharification. The pretreatment was observed to reduce lignin content by 32%, while the hemicellulose was dissolved [87]. Glucose yield was around 90% when beetle-killed lodgepole pine was pretreated with SPORL using sulfuric acid and sodium bisulfite of 2.2 and 8% for 75 min at 165 °C [88]. Sodium bisulfite of 8–10% and sulfuric acid of 3.7% on spruce chip for 30 min at 180 °C was noticed to produce a cellulose conversion rate of above 90% [83]. The application of SPORL pretreatment remains limited in lignocellulose biomass pretreatment, despite its success in municipal solid waste applications. There is a need for more studies in this regard to establish the efficiency of this novel technology. The economic feasibility and environmental impacts of this technique also need to be investigated and compared with other pretreatment methods.

4.7 ORGANIC SOLVENT

Organic solvent/organosolv pretreatment is the process of lignocellulose delignification with the use of organic solvents like glycol, methanol, acetone, ethanol, ethylene glycol, 1-pentanol, methyl isobutyl ketone, tetrahydrofuran, etc., with or without a catalyst [89–91]. This technique is typically operated at a higher temperature of approximately 200 °C; however, the use of a catalyst has reduced the required temperature. Organic or inorganic bases or acid catalysts, such as sodium hydroxide [92], sulfuric acid [93], and iron chloride ($FeCl_3$) [94], can be used to support organosolv pretreatment to enhance the enzymatic hydrolysis of biomass by degrading the lignin content. This technique eliminates hemicellulose and cellulose through the cleavage of β-O-aryl bonds, α-O-aryl bonds, and 4-O-methylglucuronic acid ester bonds (solubilization) via acidolysis and/or homolytic cleavage, thereby improving the available surface area and pore sizes of cellulose [95]. During this process, lignin is isolated as solids, while carbohydrates are isolated as a syrup and subsequently recovered as by-products. Depending on the catalyst type, the primary reaction of lignin in organosolv pretreatment is related to the reactions that occur during acid or alkaline pretreatment. Apart from the main lignin reaction, lignin side bonds (primarily the Cα position) are simultaneously alkylated during alcohol organosolv pretreatment using an acidic catalyst. The capturing of the reactive cationic intermediate by the alcohol significantly subdues further lignin degradation/condensation [43]. Furthermore, some percentages of hemicelluloses are degraded and dissolved into smaller compounds [96].

Organosolv pretreatment has been investigated on different lignocellulose biomasses and has been reported as an effective method. For instance, corn stover was pretreated with aqueous ethanol and n-propylamine, and the synergistic effects of these solvents resulted in 81.7% biomass delignification and 83.2% sugar yield [97]. It was also noted that the organic amine catalysts degrade the bond that links hemicellulose and lignin, making it a promising solvent for high-lignin biomass. Organic solvent combined with 1.5% H_2SO_4 was applied to spent coffee waste during the extraction of reducing sugar and polyphenols. It was observed that the lignin was reduced by 24.4%, accompanied by a 1.27-fold increase in reducing sugar and a 7.6-fold increase in total flavonoid [98]. Table 4.3 presents the findings of some studies on the application of the organosolv pretreatment technique. It was observed that the efficiency of the process is high when a catalyst is used, as it reduces the energy required since a lower temperature is needed. The impacts of organosolv pretreatment vary with different factors, as shown in Table 4.3. When the same biomass is pretreated with different solvents/catalysts, the results are not the same. The primary advantage of this technique is the effective hydrolysis of hemicellulose and lignin, resulting in the release of pure lignin and other by-products [99]. The technique is environmentally friendly because organic solvents are less toxic. However, several limitations hinder the successful application of the method. When compared to other chemical pretreatments with inorganic acid, it is a high-cost technique because more energy is needed for solvent recovery [100]. The technique is still suffering setbacks in commercial applications due to the degree of volatility of the organic solvent. The use of low-molecular-weight alcohols was identified as a promising option for industrial applications, but it is risky considering their volatility compared to non-volatile organic solvents [101]. Therefore, there is a need to optimize pretreatment conditions, such as solvent concentration, temperature, exposure time, and catalyst type, to achieve an optimum yield, quality, and cost. Having a continuous pretreatment setup suitable for various biomass types will enhance the acceptability of the technique.

4.8 SUPERCRITICAL CARBON DIOXIDE EXPLOSION

Supercritical carbon dioxide pretreatment is a modified steam explosion method where CO_2 is used instead of atmospheric air. It is regarded as a bright lignocellulose pretreatment method. This technique utilizes elevated pressure and temperature of the fluid at supercritical conditions to enhance performance by increasing solvent penetration through enlarged fiber pores and defects [111]. A certain quantity of wet biomass is loaded in a reactor, and pressure is built in the reactor using supercritical CO_2. This pressure is maintained for a period before it is suddenly released, resulting in the fractionalization of biomass. The supercritical CO_2 then diffuses into the crystalline arrangement of cellulose and disrupts the substrate cell wall, creating access for cellulolytic enzymes during downstream pretreatment [112]. The use of CO_2 during pretreatment produces physical and chemical effects, which involve the penetration of high-pressure CO_2 into the small pores of the biomass, leading to microstructural arrangement degradation [113,114]. Furthermore, the supercritical CO_2 encourages high diffusivity and has a swelling impact on the biomass. The

TABLE 4.3
Effect of organosolv pretreatment on some lignocellulose biomass

Feedstock	Treatment conditions	Impacts	Reference
Sugarcane bagasse	Ethyl glycerol + 0.005 mol HCl for 60 min at 150 °C.	Lignin and hemicellulose removal were 67.1 and 99.3%, respectively. Glucose and xylose recovery after saccharification was 94.3 and 77.3%, respectively.	[102]
Rice husk	60% (v/v) ethanol + 0.25% (w/v) H_2SO_4 for 60 min at 180 °C.	77.5 and 86.8% lignin and xylan removal, respectively, with 88–90% glucan recovery.	[103]
Sugarcane bagasse	Ethyl glycerol + 0.005 mol NaOH for 60 min at 150 °C.	Lignin and hemicellulose removal were 90.9 and 28.8%, respectively. Glucose and xylose recovery after saccharification was 92 and 93.5%, respectively.	[102]
Eucalyptus	n-butanol + 1% v/v H_2SO_4 for 180 °C for 45 min.	87 and 80% lignin and xylan removal, with 80% glucan recovery. Enzymatic saccharification was enhanced from 18.7 to 96%.	[104]
Birch sawdust	60% ethanol + 1% (w/w) H_2SO_4 for 15 min at 200 °C.	121.4 ml/gVS of green H_2 with 17.8 and 2.64 g/l of short and medium-chain carboxylic acids were produced during acidogenic fermentation.	[105]
Birch sawdust	60% (v/v) ethanol + 1% (v/v) H_2SO_4 for 15 min at 200 °C.	The biomass solubility of 68.07% and lignin, hemicellulose, and cellulose solubility of 87.91, 98.33, and 44.16%, respectively, were recorded.	[106]
Bamboo	85% formic acid for 45 min at 145 °C.	Glucan recovery of 98%, and xylan and lignin removal of 22 and 83%, respectively, 3.8% acetic acid, and 3.56% furfural.	[107]
Palm tree empty fruit bunch	50% (v/v) ethylene glycol + 3% (v/v) NaOH at 80 °C for 30 min.	Lignin and xylan removal of 67.2 and 54%, respectively, resulted in 67.2% delignification and 90.6% glucan recovery.	[108]
Cauliflower waste	100% (v/v) methanol + 0.1 M Na-acetate for 45 min at 121 °C.	Hydrogen, β-ester, and β-ether bonds were degraded to increase the available cellulose by 304%.	[109]
Rice husk	Methanol + 12.5% (w/v) NaOH at 80 °C for 90 min.	Lignin removal was 37.3% while xylan and glucan recovery were 83.5 and 97.5%, respectively. The enzymatic degradation of xylan and glucan was 80.3 and 97.2%, respectively.	[110]

reaction of CO_2 and water produces carbonic acid, as shown in Equation 4.1 [115]. Consequently, the pH of the CO2/H2O mixture decreases to 2.9, creating a strongly acidic environment that facilitates the hydrolysis of biomass. Supercritical CO_2 hydrolysis has been shown to exhibit efficient glucan degradation [115].

$$H_2O + CO_2 \leftrightharpoons H_2CO_3 \leftrightharpoons H^+ + HCO_3^{2-} \leftrightharpoons 2H^+ + CO_3^{2-} \quad (4.1)$$

The dissolved CO_2 in water supports acid-catalyzed degradation of biomass by releasing carbonic acid, which increases the hydronium ion concentration due to the instability of the acid. The reduction in pH in this case is sufficient to promote the degradation and hydrolysis of hemicelluloses into their corresponding sugars, resulting in improved cellulose enzymatic hydrolysis as well [116]. However, the acidic solution produced does not pose a problem to the environment, as the pH of the solution increases due to CO_2 removal during the depressurization stage. During the depressurization phase, CO_2 can also serve as a detoxification material. Due to the high-pressure CO_2 required to solubilize certain compounds, it can assist in removing inhibitory materials, such as furans produced from the hydrolysate [116]. Another merit of this technique is that the biomass is totally sterilized from viruses, bacteria, pathogens, and biotoxins. The technique does not generate inhibitory materials compared to other methods and is less corrosive. The CO_2 used in the method is non-toxic, non-flammable, and inexpensive [117]. It does not generate waste and does not require a further recovery process since the CO_2 is removed during depressurization [118]. In general, CO_2 explosion pretreatment is a bright technique to enhance the efficiency and sustainability of bioenergy recovery from lignocellulose biomass. However, this technique presents some challenges that require attention for its widespread adoption at the industrial level. The process required higher pressure, which increased the energy required, posing a major setback. A techno-economic analysis of this technique is still lacking in the literature, indicating a need for additional research on the design of a pilot-scale reactor.

4.9 IONIC LIQUIDS

The use of ionic liquids like ammonium-, imidazolium-, phosphonium-, and pyridinium-based cations, together with allyl or alkyl side chains with different anions, like acetate, phosphonate, and chloride, has been investigated for their ability to break down the recalcitrant properties of lignocellulose biomass [119], as illustrated in Figure 4.2. This method cleaves the intra- and intermolecular hydrogen bonds in lignocellulose biomass through hydrogen bond formation with sugar hydroxyl groups, thereby solubilizing lignin and cellulose [120]. It can dissolve or separate the primary contents (lignin, hemicellulose, and cellulose) of the lignocellulose biomass and degrade it into value-added products. According to their ability to degrade the biomass components, the ionic liquids are classified into four groups. The first class of ionic liquid are the ones that can degrade all three constituents (e.g., N-methyl-2-hydroxyethylammonium acetate) [121], the second class can degrade lignin and hemicelluloses (e.g., 1-butyl-3-methylimidazolium chloride) [122], the third class can degrade lignin (e.g., triethylammonium hydrogen sulfate) [123], and the fourth class can selectively degrade cellulose (e.g., 1-ethyl-3-methylimidazolium-acetate, 1-allyl-3-methylimidazolium formate) [124,125]. Ionic liquids can be easily prepared with various anions and cations, resulting in both hydrophilic and hydrophobic types, depending on the expected application [126]. Cations of methylpyridinium and methylpyridinium with ethyl-, allyl-, or

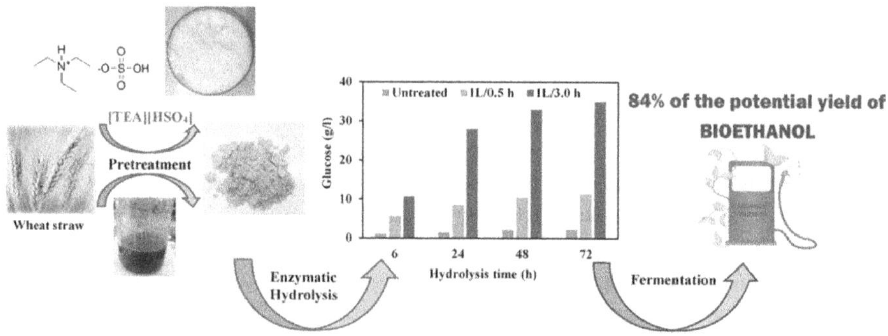

FIGURE 4.2 Ionic pretreatment of lignocellulose biomass [136].

butyl-chains are the most applied. Sulphites, halogens, phosphates, imides, sulfonates, formates, dichloroaluminates, acetates, amides, and thiocyanates are some of the most efficient anions reported [127].

Several studies on the application of ionic liquids for lignocellulose biomass pretreatment have demonstrated their potential to enhance bioenergy production. Ethanolamine acetate ([EOA] [OAc]) and cholinium lysinate ([Ch][Lys]) ionic liquids were applied to several woods before enzymatic hydrolysis and subsequent ethanol fermentation, and it was observed that 14–80% xylose and 24–84% glucose were released [120]. Triticale straw pretreatment with 1-ethyl-3-methylimidazolium acetate ([Emim]Cl) for 90 min at 150 °C removes 29% of the total lignin [128]. Ionic liquid [TEA][HSO_4] was used to pretreat poplar wood at 140 °C for 7 hrs and a 5% solid loading rate. It was observed that 70.1 and 90% of lignin and xylan, respectively, were removed, with a glucan recovery rate of 86.47%. A glucose yield of 77.55 and 100% xylose release was recorded during saccharification after enzymatic hydrolysis [129]. Combined ionic liquid of [Emim][OAc], [Emim] [Cl], and [Emim] [HSO4] with ultrasound irradiation for the pretreatment of bamboo produced 14.99% lignin and 63.99% cellulose [130].

The efficiency of ionic pretreatment depends on various parameters, including the physicochemical characteristics of the ionic liquid, the type of biomass, exposure time, feedstock moisture content, temperature, and the ratio of ionic liquid to feedstock. Viscosity is a major property that negatively impacts the process because a high-viscosity compound is released during pretreatment, significantly influencing the commercial application of this method [131]. This pretreatment technique is environmentally friendly since ionic liquids are compatible with the environment. Compared to conventional solvents, they have unique characteristics, such as low pressure, which prevents solvent loss through evaporation. They also have low flammability, high thermal stability, and recyclability, which assist in waste reduction [132]. Despite the several benefits of ionic liquid pretreatment, there are some challenges that are limiting its successful application. One of the major limitations is that the technique incurs high costs at the industrial scale. To make it feasible economically, solvent recovery of more than 97% and heat recovery of above 90%

are necessary, which is a challenge [133,134]. A high content of fresh biomass (up to 50%) and the hygroscopic nature of ionic liquids make it extremely challenging to maintain a low moisture content in lignocellulosic pretreatment [135]. The small quantity of ionic liquid remaining in the pretreated feedstock is toxic to fermentative microbes and hydrolytic enzymes. Due to these challenges, the application of ionic liquids remains primarily at the laboratory scale. Therefore, further research is required to develop an efficient approach and an economic recovery technique for recovering pretreated lignin and cellulose from ionic liquid. There is a need to investigate and utilize green and low-cost ionic liquids, as well as to incorporate conventional organic solvents into ionic liquids to form organic electrolytes for lignocellulose biomass pretreatment.

4.10 AMMONIA FIBER EXPLOSION

Ammonia fiber explosion (AFEX) is the pretreatment technique that treats lignocellulose biomass using liquid ammonia at high pressure and temperature to depolymerize the feedstock's morphological arrangement [137]. This is carried out in special equipment equipped with a liquid ammonia inflow channel and a pressure and temperature controller. During AFEX pretreatment, the feedstock and ammonia are heated to a specified temperature and held for a specific exposure time before being rapidly released through the vent valve. NH_3 disintegrates to form ammonium and hydroxide ions at high pressure, resulting in a rapid temperature increase. The rapid release of pressure results in the evaporation of NH_3, and the temperature starts to decrease. This leads to feedstock swelling, reduction in cellulose crystallinity, hemicellulose deacetylation, and degradation of lignin-carbohydrate bonds [1]. This technique removes the acetyl groups of hemicellulose, cleaves the C–O–C linkages, cleaves the lignin–carbohydrate complex bonds, and depolymerizes the cellulose contents. This has a significant influence on lignocellulose, which improves the accessibility of microbes during the downstream biorefinery [138]. The primary process parameters that influence the efficiency of this technology are exposure time, NH_3 loading, water loading, and temperature. The influence of feedstock particle size on the performance of this technique is still complicated, as some studies observed that it has a significant effect, while some reported no significant impact [10,139]. Ammonia recovery and reuse after pretreatment are necessary to enhance the process's efficiency.

AFEX pretreatment has been investigated on different lignocellulose feedstocks. For example, AFEX pretreatment of agave residue improved the biofuel generation by 48% [140]. Ethanol yield from switchgrass was increased by 2.5-fold when AFEX pretreatment was applied using a 1:1 NH3 loading rate at 100 °C for 5 min with 80% moisture content [141]. Wheat straw pretreated with 0.70% ammonia at 105 °C was observed to achieve an efficiency of 99.3%. It was observed that the methane yield was improved by 31.9% compared to the untreated feedstock [142]. Sugarcane bagasse was pretreated with 10% v/v aqueous ammonia at different temperature ranges (50–70 °C), exposure times of 12 and 24 hrs, and ethanol concentrations of 5, 25, and 50% v/v. It was observed that an optimum yield improvement of 183% was achieved when 10% ammonia was combined with 50% ethanol and applied for 12 hrs at 70 °C [143]. This technique has been reported to have the potential to improve the biobased fuels and chemicals from lignocellulose biomass. It is a dry-to-dry method that does not require washing or detoxification after pretreatment, generates low inhibitory compounds, no sugar loss, retains high hemicellulose and cellulose portions, and requires short exposure time and moderate temperature [1]. One of the primary limitations of this technique is low efficiency when applied to biomass with high lignin content [10]. The high cost of ammonia and the requirement for a controlled environment with special equipment make the investment and operation costs higher. Ammonia recovery and reuse remain significant limitations to this process.

4.11 DEEP EUTECTIC SOLVENTS

Deep eutectic solvent (DES) pretreatment is a green and alternative approach to biomass pretreatment compared to ionic liquid treatment. An eutectic mixture is a mixture of two or more components, i.e., Brønsted or Lewis acids and bases, with different cationic and/or anionic species and other molecular compounds. They can behave as hydrogen bond donors (HBD) and hydrogen bond acceptors (HBA) species, reducing the lattice energy of the process, and exhibit a melting point below that of the individual components due to the strong hydrogen-bonding interaction between HBA and HBD [144]. DES pretreatment has the capacity to solubilize lignin under mild conditions. During biomass pretreatment, partial depolymerization and repolymerization reactions may occur during lignin removal from biomass. The DES pretreatment degradation of proton-induced links in lignin can be attributed to the protonation of benzene carbon (Cα) and the subsequent dehydration of reactive Cα cations. Lignin concentrations can be produced through the combination of intermediates generated by the degradation of β-O-4 aryl ether bonds [145]. In this pretreatment technique, the degradation of lignin–carbohydrate complexes results in the release of hemicellulose, thereby enhancing cellulose accessibility and the hydrolysis rate. The performance of DES pretreatment on hemicellulose depends on the type of DES, the type of biomass, and the pretreatment conditions [146]. The impacts of DES pretreatment on cellulose are through the degradation of old bonds (initial intermolecular hydrogen bonds in cellulose) and the generation of new links (hydrogen bonds produced between -OH and DESs) [147]. The primary alterations during pretreatment are the elimination of cellulose crystal arrangement or a change in the crystallinity index. For instance, the removal of amorphous contents will improve the cellulose crystallinity index. Lignin and hemicellulose are amorphous components, and DES pretreatment removes them to enhance the feedstock's enzyme-linking sites, thereby increasing the enzymatic hydrolysis rate of cellulose [148].

DES pretreatment has been experimented with different lignocellulose biomass and was reported to improve the efficiency of downstream bioprocessing [145,148]. DES pretreatment using choline chloride and glycerol at a 1:1 ratio for 1 hr at 100 °C with a 1:2 solid:liquid ratio. It was reported that DES pretreatment significantly disrupted the morphological arrangement, improved the crystallinity index, and functional group, resulting in a 226% increase in biomethane compared to the untreated feedstock [146]. DES pretreatment on corn straw with urea and ammonium thiocyanate for 60 min at 100 °C was observed to improve biomethane yield by 48% when a 1:2 solid-to-liquid ratio was used [149]. DES solvents are easy to prepare because they do not require a complex purification process, are cost-effective, and are low in toxicity, with high biodegradability [150]. It also has some unique properties, such as easy miscibility with water and other solvents, high polarity, and the ability to selectively separate lignin contents from the biomass under mild conditions. The technique does not require a high-energy demand, as higher pressure and temperature are not necessary, and it reduces the risk of feedstock degradation. The technology generates low-inhibitory compounds and solvent recovery, allowing for reuse. However, there are some limitations that hinder its acceptability and application. These include the high cost of DES synthesis, complex design and optimization, low feedstock penetration, poor compatibility with downstream bioprocesses, and scale-up limitations due to the requirement for huge amounts of solvents, handling, and disposal, as well as low efficiency of solvent recovery and recycling [7].

4.12 ENVIRONMENTAL AND SOCIAL IMPACTS

Chemical pretreatment techniques are one of the widely studied pretreatment methods reported to have the ability to break down lignocellulose feedstocks. Efficiency and economy depend on the materials and the chemicals considered. However, this pretreatment is facing numerous challenges related to the environment and society. Some of the chemicals required for pretreatment are toxic or release harmful compounds that are dangerous to the environment. Replacing some of these chemicals with green solvents or modifying the process is required in addition to the circular economy to achieve an environmentally benign and sustainable chemical pretreatment [151,152]. Chemical pretreatment with acids, alkali, etc., requires special equipment due to the corrosiveness of the chemicals and their hazardous nature to human health [153]. This technique has a higher likelihood of generating inhibitory compounds that can have a detrimental impact on the downstream process. The operation cost of this method could be high, considering the chemical cost and the need for high-pressure and high-temperature conditions in some of the methods. Chemical neutralization after pretreatment requires a large quantity of water, and the recycling process may make the technology uneconomical [51].

4.13 CONCLUSION

Chemical pretreatment is an effective method for degrading the recalcitrant properties of lignocellulose biomass. Acid, alkali, oxidative, organic, supercritical, ammonia, SPORL, and others are some of the conventional techniques. Recent studies have also presented ionic and DESs as green chemical pretreatments. It can effectively degrade lignin, making cellulose available for downstream processing. This technique can be used as a pretreatment or for fractionalization and will play a significant role in lignocellulose biorefinery. Some of the chemical pretreatments are still at the laboratory stage, while some are already at the prototype stage, with only a few being tested on an industrial scale. Effective and efficient biomass degradation can be achieved with the proper selection of chemicals, concentration, exposure time, and temperature. The release of inhibitory compounds, the requirement for specialized equipment due to corrosion, the need to neutralize the feedstock, chemical recovery, and the high cost of chemicals are some of the challenges that limit the application of chemical pretreatment. Further studies are needed to develop economical, sustainable, environmentally benign, and upscale chemical pretreatment methods that can ensure the generation of high-quality biofuels without compromising the products. The identification of deeper scientific analysis and knowledge of process-driven mechanisms will assist in designing a tailor-made and competitive technique for future biorefinery projects. Chemical pretreatment techniques can be enhanced by varying various input parameters of the treatment conditions and the reactor setup.

REFERENCES

[1] B. Kumar, N. Bhardwaj, K. Agrawal, V. Chaturvedi, P. Verma, Current perspective on pretreatment technologies using lignocellulosic biomass: An emerging biorefinery concept, Fuel Process Technol 199 (2020) 106244. https://doi.org/10.1016/J.FUPROC.2019.106244.

[2] A.V.S.L. Sai Bharadwaj, S. Dev, J. Zhuang, Y. Wang, C.G. Yoo, B.H. Jeon, S. Aggarwal, S.H. Park, T.H. Kim, Review of chemical pretreatment of lignocellulosic biomass using low-liquid and low-chemical catalysts for effective bioconversion, Bioresour Technol 368 (2023) 128339. https://doi.org/10.1016/J.BIORTECH.2022.128339.

[3] X. Meng, A. Parikh, B. Seemala, R. Kumar, Y. Pu, P. Christopher, C.E. Wyman, C.M. Cai, A.J. Ragauskas, Chemical transformations of poplar lignin during cosolvent enhanced lignocellulosic fractionation process, ACS Sustain Chem Eng 6 (2018) 8711–8718. https://doi.org/10.1021/ACSSUSCHEMENG.8B01028/SUPPL_FILE/SC8B01028_SI_001.PDF.

[4] K.O. Olatunji, N.A. Ahmed, O. Ogunkunle, Optimization of biogas yield from lignocellulosic materials with different pretreatment methods: A review, Biotechnol Biofuels 14 (2021) 1–34. https://doi.org/10.1186/S13068-021-02012-X.

[5] S. Banu Jamaldheen, M.B. Kurade, B. Basak, C.G. Yoo, K.K. Oh, B.H. Jeon, T.H. Kim, A review on physico-chemical delignification as a pretreatment of lignocellulosic biomass for enhanced bioconversion, Bioresour Technol 346 (2022) 126591. https://doi.org/10.1016/J.BIORTECH.2021.126591.

[6] X. Li, Y. Shi, W. Kong, J. Wei, W. Song, S. Wang, Improving enzymatic hydrolysis of lignocellulosic biomass by bio-coordinated physicochemical pretreatment—A review, Energy Rep 8 (2022) 696–709. https://doi.org/10.1016/J.EGYR.2021.12.015.

[7] R.S. Abolore, S. Jaiswal, A.K. Jaiswal, Green and sustainable pretreatment methods for cellulose extraction from lignocellulosic biomass and its applications: A review, Carbohydr Polym Technol Appl 7 (2024) 100396. https://doi.org/10.1016/J.CARPTA.2023.100396.

[8] A.R. Mankar, A. Pandey, A. Modak, K.K. Pant, Pretreatment of lignocellulosic biomass: A review on recent advances, Bioresour Technol 334 (2021) 125235. https://doi.org/10.1016/J.BIORTECH.2021.125235.

[9] R. Roy, M.S. Rahman, D.E. Raynie, Recent advances of greener pretreatment technologies of lignocellulose, Curr Res Green Sustainable Chem 3 (2020) 100035. https://doi.org/10.1016/J.CRGSC.2020.100035.

[10] D. Kumari, R. Singh, Pretreatment of lignocellulosic wastes for biofuel production: A critical review, Renew Sustain Energy Rev 90 (2018) 877–891. https://doi.org/10.1016/J.RSER.2018.03.111.

[11] X. Bai, P.A. Lant, P.D. Jensen, S. Astals, S. Pratt, Enhanced methane production from algal digestion using free nitrous acid pre-treatment, Renew Energy 88 (2016) 383–390. https://doi.org/10.1016/J.RENENE.2015.11.054.
[12] K.O. Olatunji, D.M. Madyira, J.O. Amos, Sustainable enhancement of biogas and methane yield of macroalgae biomass using different pretreatment techniques: A mini-review, Energy Environ 35 (2024) 1050–1088. https://doi.org/10.1177/0958305X231193869.
[13] W. Den, V.K. Sharma, M. Lee, G. Nadadur, R.S. Varma, Lignocellulosic biomass transformations via greener oxidative pretreatment processes: Access to energy and value added chemicals, Front Chem 6 (2018) 337170. https://doi.org/10.3389/FCHEM.2018.00141/XML/NLM.
[14] A. Lorenci Woiciechowski, C.J. Dalmas Neto, L. Porto de Souza Vandenberghe, D.P. de Carvalho Neto, A.C. Novak Sydney, L.A.J. Letti, S.G. Karp, L.A. Zevallos Torres, C.R. Soccol, Lignocellulosic biomass: Acid and alkaline pretreatments and their effects on biomass recalcitrance—Conventional processing and recent advances, Bioresour Technol 304 (2020) 122848. https://doi.org/10.1016/J.BIORTECH.2020.122848.
[15] Y.H. Jung, K.H. Kim, Acidic Pretreatment, Pretreatment of Biomass: Processes and Technologies (2015) 27–50. https://doi.org/10.1016/B978-0-12-800080-9.00003-7.
[16] L.J. Jönsson, C. Martín, Pretreatment of lignocellulose: Formation of inhibitory by-products and strategies for minimizing their effects, Bioresour Technol 199 (2016) 103–112. https://doi.org/10.1016/J.BIORTECH.2015.10.009.
[17] K.S. Boes, R.H. Narron, Y. Chen, S. Park, N.R. Vinueza, Characterization of biofuel refinery byproduct via selective electrospray ionization tandem mass spectrometry, Fuel 188 (2017) 190–196. https://doi.org/10.1016/J.FUEL.2016.10.016.
[18] J. He, C. Huang, C. Lai, C. Huang, M. Li, Y. Pu, A.J. Ragauskas, Q. Yong, The effect of lignin degradation products on the generation of pseudo-lignin during dilute acid pretreatment, Ind Crops Prod 146 (2020) 112205. https://doi.org/10.1016/J.INDCROP.2020.112205.
[19] B. Liu, L. Liu, B. Deng, C. Huang, J. Zhu, L. Liang, X. He, Y. Wei, C. Qin, C. Liang, S. Liu, S. Yao, Application and prospect of organic acid pretreatment in lignocellulosic biomass separation: A review, Int J Biol Macromol 222 (2022) 1400–1413. https://doi.org/10.1016/J.IJBIOMAC.2022.09.270.
[20] C.Z. Chen, M.F. Li, Y.Y. Wu, R.C. Sun, Integration of ambient formic acid process and alkaline hydrogen peroxide post-treatment of furfural residue to enhance enzymatic hydrolysis, Ind Eng Chem Res 53 (2014) 12935–12942. https://doi.org/10.1021/IE502303S/ASSET/IMAGES/MEDIUM/IE-2014-02303S_0007.GIF.
[21] H.S. Ghaziaskar, S. Afsari, M. Rezayat, H. Rastegari, Quaternary solubility of acetic acid, diacetin and triacetin in supercritical carbon dioxide, J Supercrit Fluids 119 (2017) 52–57. https://doi.org/10.1016/J.SUPFLU.2016.09.005.
[22] G.Y. S Mtui, Oxalic acid pretreatment, fungal enzymatic saccharification and fermentation of maize residues to ethanol, Afr J Biotechnol 11 (2012) 843–851. https://doi.org/10.5897/AJB11.3032.
[23] T.M. Lacerda, M.D. Zambon, E. Frollini, Oxalic acid as a catalyst for the hydrolysis of sisal pulp, Ind Crops Prod 71 (2015) 163–172. https://doi.org/10.1016/J.INDCROP.2015.03.072.
[24] J. Moncada, J. Tamayo, C.A. Cardona, Evolution from biofuels to integrated biorefineries: Techno-economic and environmental assessment of oil palm in Colombia, J Clean Prod 81 (2014) 51–59. https://doi.org/10.1016/J.JCLEPRO.2014.06.021.
[25] J.C. Martínez-Patiño, T.A. Lu-Chau, B. Gullón, E. Ruiz, I. Romero, E. Castro, J.M. Lema, Application of a combined fungal and diluted acid pretreatment on olive tree biomass, Ind Crops Prod 121 (2018) 10–17. https://doi.org/10.1016/J.INDCROP.2018.04.078.

[26] R.R. Gonzales, P. Sivagurunathan, S.H. Kim, Effect of severity on dilute acid pretreatment of lignocellulosic biomass and the following hydrogen fermentation, Int J Hydrogen Energy 41 (2016) 21678–21684. https://doi.org/10.1016/J.IJHYDENE.2016.06.198.

[27] X. Tao, J. Li, P. Zhang, M. Nabi, S. Jin, F. Li, S. Wang, J. Ye, Reinforced acid-pretreatment of Triarrhena lutarioriparia to accelerate its enzymatic hydrolysis, Int J Hydrogen Energy 42 (2017) 18301–18308. https://doi.org/10.1016/J.IJHYDENE.2017.04.149.

[28] K.O. Olatunji, D.M. Madyira, Effect of acidic pretreatment on the microstructural arrangement and anaerobic digestion of Arachis hypogea shells; and process parameters optimization using response surface methodology, Heliyon 9 (2023) e15145. https://doi.org/10.1016/J.HELIYON.2023.E15145.

[29] K. Jaisamut, L. Paulová, P. Patáková, S. Kotúčová, M. Rychtera, Effect of sodium sulfite on acid pretreatment of wheat straw with respect to its final conversion to ethanol, Biomass Bioenergy 95 (2016) 1–7. https://doi.org/10.1016/J.BIOMBIOE.2016.08.022.

[30] P. Amnuaycheewa, R. Hengaroonprasan, K. Rattanaporn, S. Kirdponpattara, K. Cheenkachorn, M. Sriariyanun, Enhancing enzymatic hydrolysis and biogas production from rice straw by pretreatment with organic acids, Ind Crops Prod 87 (2016) 247–254. https://doi.org/10.1016/J.INDCROP.2016.04.069.

[31] S.O. Dahunsi, A.T. Adesulu-Dahunsi, C.O. Osueke, A.I. Lawal, T.M.A. Olayanju, J.O. Ojediran, J.O. Izebere, Biogas generation from Sorghum bicolor stalk: Effect of pretreatment methods and economic feasibility, Energy Rep 5 (2019) 584–593. https://doi.org/10.1016/j.egyr.2019.04.002.

[32] S.M. de Vasconcelos, A.M.P. Santos, G.J.M. Rocha, A.M. Souto-Maior, Diluted phosphoric acid pretreatment for production of fermentable sugars in a sugarcane-based biorefinery, Bioresour Technol 135 (2013) 46–52. https://doi.org/10.1016/J.BIORTECH.2012.10.083.

[33] M. Kuglarz, M. Alvarado-Morales, K. Dąbkowska, I. Angelidaki, Integrated production of cellulosic bioethanol and succinic acid from rapeseed straw after dilute-acid pretreatment, Bioresour Technol 265 (2018) 191–199. https://doi.org/10.1016/J.BIORTECH.2018.05.099.

[34] A. Dasgupta, M.K. Chandel, Enhancement of biogas production from organic fraction of municipal solid waste using acid pretreatment, SN Appl Sci 2 (2020) 1–14. https://doi.org/10.1007/S42452-020-03213-Z/TABLES/5.

[35] W.W. Kululo, N.G. Habtu, M.K. Abera, Z.B. Sendekie, S.W. Fanta, T.A. Yemata, Advances in various pretreatment strategies of lignocellulosic substrates for the production of bioethanol: A comprehensive review, Discover Appl Sci 7 (2025) 1–24. https://doi.org/10.1007/S42452-025-06748-1.

[36] H. Xu, B. Li, X. Mu, Review of alkali-based pretreatment to enhance enzymatic saccharification for lignocellulosic biomass conversion, Ind Eng Chem Res 55 (2016) 8691–8705. https://doi.org/10.1021/acs.iecr.6b01907.

[37] N. Lahboubi, F. Karouach, M. Bakraoui, Y. El Gnaoui, A. Essamri, H. El Bari, Effect of alkali-NaOH pretreatment on methane production from anaerobic digestion of date palm waste, Ecol Eng Environ Technol 23 (2022) 78–89. https://doi.org/10.12912/27197050/144846.

[38] G. Bali, X. Meng, J.I. Deneff, Q. Sun, A.J. Ragauskas, The effect of alkaline pretreatment methods on cellulose structure and accessibility, ChemSusChem 8 (2015) 275–279. https://doi.org/10.1002/CSSC.201402752.

[39] E. van der Pol, R. Bakker, A. van Zeeland, D. Sanchez Garcia, A. Punt, G. Eggink, Analysis of by-product formation and sugar monomerization in sugarcane bagasse pretreated at pilot plant scale: Differences between autohydrolysis, alkaline and acid

pretreatment, Bioresour Technol 181 (2015) 114–123. https://doi.org/10.1016/J.BIORTECH.2015.01.033.
[40] Anu, A. Kumar, A. Rapoport, G. Kunze, S. Kumar, D. Singh, B. Singh, Multifarious pretreatment strategies for the lignocellulosic substrates for the generation of renewable and sustainable biofuels: A review, Renew Energy 160 (2020) 1228–1252. https://doi.org/10.1016/J.RENENE.2020.07.031.
[41] G.D. Saratale, M.K. Oh, Improving alkaline pretreatment method for preparation of whole rice waste biomass feedstock and bioethanol production, RSC Adv 5 (2015) 97171–97179. https://doi.org/10.1039/C5RA17797A.
[42] J. Li, Y.N. Zha, H.M. Wang, J.N. Tian, Q.X. Hou, Advances in lignin chemistry during pulping and bleaching, Ind Crops Prod 229 (2025) 121004. https://doi.org/10.1016/J.INDCROP.2025.121004.
[43] C. Huang, X. Jiang, X. Shen, J. Hu, W. Tang, X. Wu, A. Ragauskas, H. Jameel, X. Meng, Q. Yong, Lignin-enzyme interaction: A roadblock for efficient enzymatic hydrolysis of lignocellulosics, Renew Sustain Energy Rev 154 (2022) 111822. https://doi.org/10.1016/J.RSER.2021.111822.
[44] J.S. Kim, Y.Y. Lee, T.H. Kim, A review on alkaline pretreatment technology for bioconversion of lignocellulosic biomass, Bioresour Technol 199 (2016) 42–48. https://doi.org/10.1016/J.BIORTECH.2015.08.085.
[45] D.M. Madyira, K.O. Olatunji, Investigating the influence of different pretreatment methods on the morphological structure of Arachis hypogea shells, Mater Today Proc 105 (2024) 72–77. https://doi.org/10.1016/J.MATPR.2023.07.289.
[46] L. Li, C. Chen, R. Zhang, Y. He, W. Wang, G. Liu, Pretreatment of corn stover for methane production with the combination of potassium hydroxide and calcium hydroxide, Energy Fuels 29 (2015) 5841–5846. https://doi.org/10.1021/ACS.ENERGYFUELS.5B01170.
[47] P. Mutrakulcharoen, P. Pornwongthong, A. Sahithi, T. Phusantisampan, A. Tawai, M. Sriariyanun, Improvement of potassium permanganate pretreatment by enzymatic saccharification of rice straw for production of biofuels, 2021. https://doi.org/10.1051/e3sconf/202130202013.
[48] H. Zabed, J.N. Sahu, A.N. Boyce, G. Faruq, Fuel ethanol production from lignocellulosic biomass: An overview on feedstocks and technological approaches, Renew Sustain Energy Rev 66 (2016) 751–774. https://doi.org/10.1016/J.RSER.2016.08.038.
[49] D.M. Madyira, K.O. Olatunji, Alkali pretreatment of lignocellulose feedstock improves morphological structure and biomethane yield, Sustainability 17 (2025) 534. https://doi.org/10.3390/SU17020534.
[50] R. Chandra, H. Takeuchi, T. Hasegawa, Methane production from lignocellulosic agricultural crop wastes: A review in context to second generation of biofuel production, Renew Sustain Energy Rev 16 (2012) 1462–1476. https://doi.org/10.1016/j.rser.2011.11.035.
[51] S. Meenakshisundaram, A. Fayeulle, E. Léonard, C. Ceballos, X. Liu, A. Pauss, Combined biological and chemical/physicochemical pretreatment methods of lignocellulosic biomass for bioethanol and biomethane energy production—A Review, Appl Microbiol 2 (2022) 716–734. https://doi.org/10.3390/APPLMICROBIOL2040055.
[52] N.R. Baral, A. Shah, Comparative techno-economic analysis of steam explosion, dilute sulfuric acid, ammonia fiber explosion and biological pretreatments of corn stover, Bioresour Technol 232 (2017) 331–343. https://doi.org/10.1016/J.BIORTECH.2017.02.068.
[53] G. Mancini, S. Papirio, P.N.L. Lens, G. Esposito, Increased biogas production from wheat straw by chemical pretreatments, Renew Energy 119 (2018) 608–614. https://doi.org/10.1016/J.RENENE.2017.12.045.

[54] J. Zhang, C. Kong, M. Yang, L. Zang, Comparison of calcium oxide and calcium peroxide pretreatments of wheat straw for improving biohydrogen production, ACS Omega 5 (2020) 9151–9161. https://doi.org/10.1021/ACSOMEGA.9B04368/ASSET/IMAGES/LARGE/AO9B04368_0005.JPEG.
[55] R.R. Gonzales, P. Sivagurunathan, S.H. Kim, Effect of severity on dilute acid pretreatment of lignocellulosic biomass and the following hydrogen fermentation, Int J Hydrogen Energy 41 (2016) 21678–21684. https://doi.org/10.1016/J.IJHYDENE.2016.06.198.
[56] K.O. Olatunji, D.M. Madyira, N.A. Ahmed, O. Ogunkunle, Influence of alkali pretreatment on morphological structure and methane yield of Arachis hypogea shells, Biomass Convers Biorefin 14 (2024) 12143–12154. https://doi.org/10.1007/S13399-022-03271-W/METRICS.
[57] P.T. Vu, Y. Unpaprom, R. Ramaraj, Impact and significance of alkaline-oxidant pretreatment on the enzymatic digestibility of sphenoclea zeylanica for bioethanol production, Bioresour Technol 247 (2018) 125–130. https://doi.org/10.1016/J.BIORTECH.2017.09.012.
[58] Y. Jin, Z. Shi, G. Xu, H. Yang, J. Yang, A stepwise pretreatment of sugarcane bagasse by alkaline and hydroxymethyl reagent for bioethanol production, Ind Crops Prod 145 (2020) 112136. https://doi.org/10.1016/J.INDCROP.2020.112136.
[59] Zahoor, W. Wang, X. Tan, M. Imtiaz, Q. Wang, C. Miao, Z. Yuan, X. Zhuang, Rice straw pretreatment with KOH/urea for enhancing sugar yield and ethanol production at low temperature, Ind Crops Prod 170 (2021) 113776. https://doi.org/10.1016/J.INDCROP.2021.113776.
[60] Z. Yuan, Y. Wen, G. Li, Production of bioethanol and value added compounds from wheat straw through combined alkaline/alkaline-peroxide pretreatment, Bioresour Technol 259 (2018) 228–236. https://doi.org/10.1016/J.BIORTECH.2018.03.044.
[61] K.O. Olatunji, D.M. Madyira, Optimization of biomethane yield of Xyris capensis grass using oxidative pretreatment, Energies 16 (2023) 3977. https://doi.org/10.3390/EN16103977.
[62] F. Monlau, A. Barakat, E. Trably, C. Dumas, J.P. Steyer, H. Carrère, Lignocellulosic materials into biohydrogen and biomethane: Impact of structural features and pretreatment, Crit Rev Environ Sci Technol 43 (2013) 260–322. https://doi.org/10.1080/10643389.2011.604258.
[63] N. Das, P.K. Jena, D. Padhi, M. Kumar Mohanty, G. Sahoo, A comprehensive review of characterization, pretreatment and its applications on different lignocellulosic biomass for bioethanol production, Biomass Convers Biorefin 2021 (2021) 1–25. https://doi.org/10.1007/S13399-021-01294-3.
[64] Z. Zhou, D. Ouyang, D. Liu, X. Zhao, Oxidative pretreatment of lignocellulosic biomass for enzymatic hydrolysis: Progress and challenges, Bioresour Technol 367 (2023) 128208. https://doi.org/10.1016/J.BIORTECH.2022.128208.
[65] M.M.I. Sheikh, C.H. Kim, H.H. Park, H.G. Nam, G.S. Lee, H.S. Jo, J.Y. Lee, J.W. Kim, A synergistic effect of pretreatment on cell wall structural changes in barley straw (Hordeum vulgare L.) for efficient bioethanol production, J Sci Food Agric 95 (2015) 843–850. https://doi.org/10.1002/JSFA.7004.
[66] Z. Song, G. Yang, Y. Guo, T. Zhang, Comparison of two chemical pretreatments of rice straw for biogas production by anaerobic digestion, Bioresources 7 (2012). https://doi.org/10.15376/biores.7.3.3223-3236.
[67] A. Morone, T. Chakrabarti, R.A. Pandey, Effect of chemical input during wet air oxidation pretreatment of rice straw in reducing biomass recalcitrance and enhancing cellulose accessibility, Korean J Chem Eng 35 (2018) 2403–2412. https://doi.org/10.1007/S11814-018-0129-2.

[68] F. Almomani, R.R. Bhosale, M.A.M. Khraisheh, M. Shawaqfah, Enhancement of biogas production from agricultural wastes via pre-treatment with advanced oxidation processes, Fuel 253 (2019) 964–974. https://doi.org/10.1016/J.FUEL.2019.05.057.
[69] K.O. Olatunji, D.M. Madyira, Enhancing microstructural and biomethane yield of Xyris capensis using combined oxidative and nanoparticle additive pretreatment methods, Biomass Convers Biorefin 1 (2023) 1–15. https://doi.org/10.1007/S13399-023-04401-8/FIGURES/5.
[70] J.T.E. Lee, M.U. Khan, H. Tian, A.W.L. Ee, E.Y. Lim, Y. Dai, Y.W. Tong, B.K. Ahring, Improving methane yield of oil palm empty fruit bunches by wet oxidation pretreatment: Mesophilic and thermophilic anaerobic digestion conditions and the associated global warming potential effects, Energy Convers Manag 225 (2020) 113438. https://doi.org/10.1016/J.ENCONMAN.2020.113438.
[71] J.J. Lamb, M.H. Islam, D.R. Hjelme, B.G. Pollet, K.M. Lien, Effect of power ultrasound and Fenton reagents on the biomethane potential from steam-exploded birchwood, Ultrason Sonochem 58 (2019) 104675. https://doi.org/10.1016/J.ULTSONCH.2019.104675.
[72] R. Travaini, J. Martín-Juárez, A. Lorenzo-Hernando, S. Bolado-Rodríguez, Ozonolysis: An advantageous pretreatment for lignocellulosic biomass revisited, Bioresour Technol 199 (2016) 2–12. https://doi.org/10.1016/J.BIORTECH.2015.08.143.
[73] A.R. Mulakhudair, J. Hanotu, W. Zimmerman, Exploiting ozonolysis-microbe synergy for biomass processing: Application in lignocellulosic biomass pretreatment, Biomass Bioenergy 105 (2017) 147–154. https://doi.org/10.1016/J.BIOMBIOE.2017.06.018.
[74] K. Tamilarasan, J.R. Banu, M.D. Kumar, G. Sakthinathan, J.H. Park, Influence of mild-ozone assisted disperser pretreatment on the enhanced biogas generation and biodegradability of green marine macroalgae, Front Energy Res 7 (2019) 89. https://doi.org/10.3389/FENRG.2019.00089/BIBTEX.
[75] C.A. Prado, F.A.F. Antunes, T.M. Rocha, S. Sánchez-Muñoz, F.G. Barbosa, R. Terán-Hilares, M.M. Cruz-Santos, G.L. Arruda, S.S. da Silva, J.C. Santos, A review on recent developments in hydrodynamic cavitation and advanced oxidative processes for pretreatment of lignocellulosic materials, Bioresour Technol 345 (2022) 126458. https://doi.org/10.1016/J.BIORTECH.2021.126458.
[76] L. Capolupo, V. Faraco, Green methods of lignocellulose pretreatment for biorefinery development, Appl Microbiol Biotechnol 100 (2016) 9451–9467. https://doi.org/10.1007/S00253-016-7884-Y.
[77] R. Patil, C. Cimon, C. Eskicioglu, V. Goud, Effect of ozonolysis and thermal pretreatment on rice straw hydrolysis for the enhancement of biomethane production, Renew Energy 179 (2021) 467–474. https://doi.org/10.1016/J.RENENE.2021.07.048.
[78] E.M. Ben'ko, O.R. Manisova, V.V. Lunin, Effect of moisture content on the interaction between lignocellulosic materials and ozone, Russian J Phys Chem A 91 (2017) 1190–1196. https://doi.org/10.1134/S0036024417070056/METRICS.
[79] D.R. Osuna-Laveaga, O. García-Depraect, R. Vallejo-Rodríguez, A. López-López, E. León-Becerril, Integrated ozonation-enzymatic hydrolysis pretreatment of sugarcane bagasse: Enhancement of sugars released to expended ozone ratio, Processes 8 (2020) 1274. https://doi.org/10.3390/PR8101274.
[80] C. Li, L. Wang, Z. Chen, Y. Li, R. Wang, X. Luo, G. Cai, Y. Li, Q. Yu, J. Lu, Ozonolysis pretreatment of maize stover: The interactive effect of sample particle size and moisture on ozonolysis process, Bioresour Technol 183 (2015) 240–247. https://doi.org/10.1016/J.BIORTECH.2015.01.042.
[81] L. Wang, Y. Luo, X. Luo, R. Wang, Y. Li, Y. Li, H. Shao, Z. Chen, Effect of deoxynivalenol detoxification by ozone treatment in wheat grains, Food Control 66 (2016) 137–144. https://doi.org/10.1016/J.FOODCONT.2016.01.038.

[82] N.S. Ab Rasid, M.M. Zainol, N.A.S. Amin, Pretreatment of agroindustry waste by ozonolysis for synthesis of biorefinery products, Refining Biomass Residues for Sustainable Energy and Bioproducts: Technology, Advances, Life Cycle Assessment, and Economics (2020) 303–336. https://doi.org/10.1016/B978-0-12-818996-2.00014-4.
[83] Y. Liang, Y. Yin, H. Zhou, Y. Tian, Enhanced enzymatic saccharification of mixed sawdust wastes: Comparison of sporl, dilute acid, formic acid, and ethanol organosolv pretreatments, J Mex Chem Soc 65 (2021) 297–307. https://doi.org/10.29356/JMCS.V65I3.1427.
[84] L. Shuai, Q. Yang, J.Y. Zhu, F.C. Lu, P.J. Weimer, J. Ralph, X.J. Pan, Comparative study of SPORL and dilute-acid pretreatments of spruce for cellulosic ethanol production, Bioresour Technol 101 (2010) 3106–3114. https://doi.org/10.1016/J.BIORTECH.2009.12.044.
[85] D.S. Zhang, Q. Yang, J.Y. Zhu, X.J. Pan, Sulfite (SPORL) pretreatment of switchgrass for enzymatic saccharification, Bioresour Technol 129 (2013) 127–134. https://doi.org/10.1016/J.BIORTECH.2012.11.031.
[86] X. Liu, M. Du, J. Yang, Y. Wu, Q. Xu, D. Wang, Q. Yang, G. Yang, X. Li, Sulfite serving as a pretreatment method for alkaline fermentation to enhance short-chain fatty acid production from waste activated sludge, Chem Eng J 385 (2020) 123991. https://doi.org/10.1016/J.CEJ.2019.123991.
[87] M. Galbe, O. Wallberg, Pretreatment for biorefineries: A review of common methods for efficient utilisation of lignocellulosic materials, Biotechnol Biofuels 12 (2019) 1–26. https://doi.org/10.1186/S13068-019-1634-1.
[88] H. Zhou, J.Y. Zhu, R. Gleisner, X. Qiu, E. Horn, J. Negrón, Pilot-scale demonstration of SPORL for bioconversion of lodgepole pine to bioethanol and lignosulfonate, Holzforschung 70 (2016) 21–30. https://doi.org/10.1515/HF-2014-0332.
[89] A. Smit, W. Huijgen, Effective fractionation of lignocellulose in herbaceous biomass and hardwood using a mild acetone organosolv process, Green Chem 19 (2017) 5505–5514. https://doi.org/10.1039/C7GC02379K.
[90] B. Seemala, X. Meng, A. Parikh, N. Nagane, R. Kumar, C.E. Wyman, A. Ragauskas, P. Christopher, C.M. Cai, Hybrid catalytic biorefining of hardwood biomass to methylated furans and depolymerized technical lignin, ACS Sustain Chem Eng 6 (2018) 10587–10594. https://doi.org/10.1021/ACSSUSCHEMENG.8B01930/SUPPL_FILE/SC8B01930_SI_001.PDF.
[91] C. Dong, X. Meng, C.S. Yeung, H.Y. Tse, A.J. Ragauskas, S.Y. Leu, Diol pretreatment to fractionate a reactive lignin in lignocellulosic biomass biorefineries, Green Chem 21 (2019) 2788–2800. https://doi.org/10.1039/C9GC00596J.
[92] B.W. Koo, H.Y. Kim, N. Park, S.M. Lee, H. Yeo, I.G. Choi, Organosolv pretreatment of Liriodendron tulipifera and simultaneous saccharification and fermentation for bioethanol production, Biomass Bioenergy 35 (2011) 1833–1840. https://doi.org/10.1016/J.BIOMBIOE.2011.01.014.
[93] K. Dussan, M. Hoek, T. de Vrije, R. van de Vondervoort, P. Bonouvrie, R. Caliskan, A. Parenti, W. Zegada-Lizarazu, A. Monti, A.T. Smit, A.M. López-Contreras, Performance of mild acetone organosolv fractionation on lignocellulosic feedstocks from new cropping systems for production of advanced bioethanol, Ind Crops Prod 223 (2025) 120156. https://doi.org/10.1016/J.INDCROP.2024.120156.
[94] H. Zhang, S. Zhang, H. Yuan, G. Lyu, J. Xie, FeCl3-catalyzed ethanol pretreatment of sugarcane bagasse boosts sugar yields with low enzyme loadings and short hydrolysis time, Bioresour Technol 249 (2018) 395–401. https://doi.org/10.1016/J.BIORTECH.2017.10.053.

[95] P. Sannigrahi, A.J. Ragauskas, Fundamentals of biomass pretreatment by fractionation, Aqueous Pretreatment of Plant Biomass for Biological and Chemical Conversion to Fuels and Chemicals (2013) 201–222. https://doi.org/10.1002/9780470975831.CH10.
[96] L. Yao, C.G. Yoo, Y. Pu, X. Meng, W. Muchero, G.A. Tuskan, T.J. Tschaplinski, A.J. Ragauskas, H. Yang, Physicochemical changes of cellulose and their influences on Populus trichocarpa digestibility after different pretreatments, Bioresources 14 (2019) 9658–9676. https://doi.org/10.15376/BIORES.14.4.9658-9676.
[97] C. Tang, J. Shan, Y. Chen, L. Zhong, T. Shen, C. Zhu, H. Ying, Organic amine catalytic organosolv pretreatment of corn stover for enzymatic saccharification and high-quality lignin, Bioresour Technol 232 (2017) 222–228. https://doi.org/10.1016/J.BIORTECH.2017.02.041.
[98] R. Ravindran, C. Desmond, S. Jaiswal, A.K. Jaiswal, Optimisation of organosolv pretreatment for the extraction of polyphenols from spent coffee waste and subsequent recovery of fermentable sugars, Bioresour Technol Rep 3 (2018) 7–14. https://doi.org/10.1016/J.BITEB.2018.05.009.
[99] J.A. Ferreira, M.J. Taherzadeh, Improving the economy of lignocellulose-based biorefineries with organosolv pretreatment, Bioresour Technol 299 (2020) 122695. https://doi.org/10.1016/J.BIORTECH.2019.122695.
[100] A. Rodrigues Gurgel da Silva, M. Errico, B.G. Rong, Techno-economic analysis of organosolv pretreatment process from lignocellulosic biomass, Clean Technol Environ Policy 20 (2018) 1401–1412. https://doi.org/10.1007/S10098-017-1389-Y/FIGURES/6.
[101] C. Martín, J. Puls, A. Schreiber, B. Saake, Optimization of sulfuric acid-assisted glycerol pretreatment of sugarcane bagasse, Holzforschung 67 (2013) 523–530. https://doi.org/10.1515/HF-2012-0179/MACHINEREADABLECITATION/RIS.
[102] W. Wei, B. Wang, X. Wang, R. Ling, Y. Jin, Comparison of acid and alkali catalyzed ethylene glycol organosolv pretreatment for sugar production from bagasse, Bioresour Technol 320 (2021) 124293. https://doi.org/10.1016/J.BIORTECH.2020.124293.
[103] T.H. Kim, H.J. Ryu, K.K. Oh, Improvement of organosolv fractionation performance for rice husk through a low acid-catalyzation, Energies 12 (2019) 1800. https://doi.org/10.3390/EN12091800.
[104] Q. Schmetz, H. Teramura, K. Morita, T. Oshima, A. Richel, C. Ogino, A. Kondo, Versatility of a dilute acid/butanol pretreatment investigated on various lignocellulosic biomasses to produce lignin, monosaccharides and cellulose in distinct phases, ACS Sustain Chem Eng 7 (2019) 11069–11079. https://doi.org/10.1021/ACSSUSCHEMENG.8B05841/SUPPL_FILE/SC8B05841_SI_002.PDF.
[105] O. Sarkar, U. Rova, P. Christakopoulos, L. Matsakas, Organosolv pretreated birch sawdust for the production of green hydrogen and renewable chemicals in an integrated biorefinery approach, Bioresour Technol 344 (2022) 126164. https://doi.org/10.1016/J.BIORTECH.2021.126164.
[106] M. Monção, K. Hrůzová, U. Rova, L. Matsakas, P. Christakopoulos, Organosolv fractionation of birch sawdust: Establishing a lignin-first biorefinery, Molecules 26 (2021) 6754. https://doi.org/10.3390/MOLECULES26216754.
[107] Y. Zhang, Q. Hou, Y. Fu, C. Xu, A.I. Smeds, S. Willför, Z. Wang, Z. Li, M. Qin, One-step fractionation of the main components of bamboo by formic acid-based organosolv process under pressure, J Wood Chem Technol 38 (2018) 170–182. https://doi.org/10.1080/02773813.2017.1388823.
[108] D.W.K. Chin, S. Lim, Y.L. Pang, L.K. Leong, C.H. Lim, Investigation of organosolv pretreatment to natural microbial-degraded empty fruit bunch for sugar based substrate recovery, Energy Procedia 158 (2019) 1065–1071. https://doi.org/10.1016/J.EGYPRO.2019.01.258.

[109] S. Majumdar, B. Goswami, A. Chakraborty, D.K. Bhattacharyya, J. Bhowal, Effect of pretreatment with organic solvent on enzymatic digestibility of cauliflower wastes, Prep Biochem Biotechnol 49 (2019) 935–948. https://doi.org/10.1080/10826068.2019.1650374.

[110] W. Yuan, Z. Gong, G. Wang, W. Zhou, Y. Liu, X. Wang, M. Zhao, Alkaline organosolv pretreatment of corn stover for enhancing the enzymatic digestibility, Bioresour Technol 265 (2018) 464–470. https://doi.org/10.1016/J.BIORTECH.2018.06.038.

[111] E.L.N. Escobar, T.A. da Silva, C.L. Pirich, M.L. Corazza, L. Pereira Ramos, Supercritical fluids: A promising technique for biomass pretreatment and fractionation, Front Bioeng Biotechnol 8 (2020) 524832. https://doi.org/10.3389/FBIOE.2020.00252/XML/NLM.

[112] R. Ravindran, A.K. Jaiswal, A comprehensive review on pre-treatment strategy for lignocellulosic food industry waste: Challenges and opportunities, Bioresour Technol 199 (2016) 92–102. https://doi.org/10.1016/J.BIORTECH.2015.07.106.

[113] E. Moharreri, T. Jafari, S.L. Suib, N. Srinivasan, A.F. Ghobadi, L.K. Ju, J.R. Elliott, Improved understanding of CO2-water pretreatment of guayule biomass by high solids ratio experiments, rapid physical expansion, and examination of textural properties, Ind Eng Chem Res 56 (2017) 645–652. https://doi.org/10.1021/ACS.IECR.6B03318.

[114] A. Toscan, A.R.C. Morais, S.M. Paixão, L. Alves, J. Andreaus, M. Camassola, A.J.P. Dillon, R.M. Lukasik, High-pressure carbon dioxide/water pre-treatment of sugarcane bagasse and elephant grass: Assessment of the effect of biomass composition on process efficiency, Bioresour Technol 224 (2017) 639–647. https://doi.org/10.1016/J.BIORTECH.2016.11.101.

[115] M. Sasaki, K. Ohsawa, Hydrolysis of lignocellulosic biomass in hot-compressed water with supercritical carbon dioxide, ACS Omega 6 (2021) 14252–14259. https://doi.org/10.1021/ACSOMEGA.1C01026/ASSET/IMAGES/MEDIUM/AO1C01026_M003.GIF.

[116] A.R.C. Morais, A.M. Da Costa Lopes, R. Bogel-Łukasik, Carbon dioxide in biomass processing: Contributions to the green biorefinery concept, Chem Rev 115 (2015) 3–27. https://doi.org/10.1021/CR500330Z/ASSET/IMAGES/LARGE/CR-2014-00330Z_0004.JPEG.

[117] F.M. Relvas, A.R.C. Morais, R. Bogel-Lukasik, Selective hydrolysis of wheat straw hemicellulose using high-pressure CO2 as catalyst, RSC Adv 5 (2015) 73935–73944. https://doi.org/10.1039/C5RA14632A.

[118] S. Haghighi Mood, A. Hossein Golfeshan, M. Tabatabaei, G. Salehi Jouzani, G.H. Najafi, M. Gholami, M. Ardjmand, Lignocellulosic biomass to bioethanol, a comprehensive review with a focus on pretreatment, Renew Sustain Energy Rev 27 (2013) 77–93. https://doi.org/10.1016/J.RSER.2013.06.033.

[119] Z. Usmani, M. Sharma, P. Gupta, Y. Karpichev, N. Gathergood, R. Bhat, V.K. Gupta, Ionic liquid based pretreatment of lignocellulosic biomass for enhanced bioconversion, Bioresour Technol 304 (2020) 123003. https://doi.org/10.1016/J.BIORTECH.2020.123003.

[120] L. Das, E.C. Achinivu, C.A. Barcelos, E. Sundstrom, B. Amer, E.E.K. Baidoo, B.A. Simmons, N. Sun, J.M. Gladden, Deconstruction of woody biomass via protic and aprotic ionic liquid pretreatment for ethanol production, ACS Sustain Chem Eng 9 (2021) 4422–4432. https://doi.org/10.1021/ACSSUSCHEMENG.0C07925/SUPPL_FILE/SC0C07925_SI_001.PDF.

[121] T.C. Pin, P.Y.S. Nakasu, S. Mattedi, S.C. Rabelo, A.C. Costa, Screening of protic ionic liquids for sugarcane bagasse pretreatment, Fuel 235 (2019) 1506–1514. https://doi.org/10.1016/J.FUEL.2018.08.122.

[122] H. Mahmood, M. Moniruzzaman, S. Yusup, H.M. Akil, Ionic liquid pretreatment at high solids loading: A clean approach for fabrication of renewable resource based particulate composites, Polym Compos 39 (2018) 1994–2003. https://doi.org/10.1002/PC.24159.
[123] C.L. Chambon, T.Y. Mkhize, P. Reddy, A. Brandt-Talbot, N. Deenadayalu, P.S. Fennell, J.P. Hallett, Pretreatment of south african sugarcane bagasse using a low-cost protic ionic liquid: A comparison of whole, depithed, fibrous and pith bagasse fractions, Biotechnol Biofuels 11 (2018) 1–16. https://doi.org/10.1186/S13068-018-1247-0/TABLES/2.
[124] J. Viell, N.K. Szekely, G. Mangiapia, C. Hövelmann, C. Marks, H. Frielinghaus, In operando monitoring of wood transformation during pretreatment with ionic liquids, Cellulose 27 (2020) 4889–4907. https://doi.org/10.1007/S10570-020-03119-4/FIGURES/8.
[125] P. Moyer, M.D. Smith, N. Abdoulmoumine, S.C. Chmely, J.C. Smith, L. Petridis, N. Labbé, Relationship between lignocellulosic biomass dissolution and physicochemical properties of ionic liquids composed of 3-methylimidazolium cations and carboxylate anions, Phys Chem Chem Phys 20 (2018) 2508–2516. https://doi.org/10.1039/C7CP07195G.
[126] L.T.P. Trinh, Y.J. Lee, J.W. Lee, H.J. Lee, Characterization of ionic liquid pretreatment and the bioconversion of pretreated mixed softwood biomass, Biomass Bioenergy 81 (2015) 1–8. https://doi.org/10.1016/J.BIOMBIOE.2015.05.005.
[127] J.U. Hernández-Beltrán, I.O. Hernández-De Lira, M.M. Cruz-Santos, A. Saucedo-Luevanos, F. Hernández-Terán, N. Balagurusamy, Insight into pretreatment methods of lignocellulosic biomass to increase biogas yield: Current state, challenges, and opportunities, Appl Sci 9 (2019) 3721. https://doi.org/10.3390/APP9183721.
[128] F.L. Wang, S. Li, Y.X. Sun, H.Y. Han, B.X. Zhang, B.Z. Hu, Y.F. Gao, X.M. Hu, Ionic liquids as efficient pretreatment solvents for lignocellulosic biomass, RSC Adv 7 (2017) 47990–47998. https://doi.org/10.1039/C7RA08110C.
[129] Z. Ziaei-Rad, M. Pazouki, Cost-effective ionic liquid pretreatment of poplar wood for biofuel production: An optimization study using response surface methodology, Ind Crops Prod 219 (2024) 119036. https://doi.org/10.1016/J.INDCROP.2024.119036.
[130] A.H.A. Rahim, Z. Man, A. Sarwono, N. Muhammad, A.S. Khan, W.S.W. Hamzah, N.M. Yunus, Y.A. Elsheikh, Probe sonication assisted ionic liquid treatment for rapid dissolution of lignocellulosic biomass, Cellulose 27 (2020) 2135–2148. https://doi.org/10.1007/S10570-019-02914-Y/FIGURES/16.
[131] X. Hu, L. Cheng, Z. Gu, Y. Hong, Z. Li, C. Li, Effects of ionic liquid/water mixture pretreatment on the composition, the structure and the enzymatic hydrolysis of corn stalk, Ind Crops Prod 122 (2018) 142–147. https://doi.org/10.1016/J.INDCROP.2018.05.056.
[132] P. Weerachanchai, S.S.J. Leong, M.W. Chang, C.B. Ching, J.M. Lee, Improvement of biomass properties by pretreatment with ionic liquids for bioconversion process, Bioresour Technol 111 (2012) 453–459. https://doi.org/10.1016/J.BIORTECH.2012.02.023.
[133] Q. Zhang, J. Hu, D.J. Lee, Pretreatment of biomass using ionic liquids: Research updates, Renew Energy 111 (2017) 77–84. https://doi.org/10.1016/J.RENENE.2017.03.093.
[134] N.R. Baral, A. Shah, Techno-economic analysis of cellulose dissolving ionic liquid pretreatment of lignocellulosic biomass for fermentable sugars production, Biofuels Bioprod Biorefin 10 (2016) 70–88. https://doi.org/10.1002/BBB.1622.
[135] F.J.V. Gschwend, F. Malaret, S. Shinde, A. Brandt-Talbot, J.P. Hallett, Rapid pretreatment of Miscanthus using the low-cost ionic liquid triethylammonium hydrogen sulfate at elevated temperatures, Green Chem 20 (2018) 3486–3498. https://doi.org/10.1039/C8GC00837J.

[136] Z. Ziaei-Rad, J. Fooladi, M. Pazouki, S.N. Gummadi, Lignocellulosic biomass pretreatment using low-cost ionic liquid for bioethanol production: An economically viable method for wheat straw fractionation, Biomass Bioenergy 151 (2021) 106140. https://doi.org/10.1016/J.BIOMBIOE.2021.106140.

[137] J. Zhang, W. Zhang, Z. Cai, J. Zhang, D. Guan, D. Ji, W. Gao, Effect of ammonia fiber expansion combined with NaOH pretreatment on the resource efficiency of herbaceous and woody lignocellulosic biomass, ACS Omega 7 (2022) 18761–18769. https://doi.org/10.1021/ACSOMEGA.2C01302/ASSET/IMAGES/LARGE/AO2C01302_0005.JPEG.

[138] R. Kumar, V. Strezov, H. Weldekidan, J. He, S. Singh, T. Kan, B. Dastjerdi, Lignocellulose biomass pyrolysis for bio-oil production: A review of biomass pretreatment methods for production of drop-in fuels, Renew Sustain Energy Rev 123 (2020) 109763. https://doi.org/10.1016/J.RSER.2020.109763.

[139] T.G. Bridgeman, L.I. Darvell, J.M. Jones, P.T. Williams, R. Fahmi, A.V. Bridgwater, T. Barraclough, I. Shield, N. Yates, S.C. Thain, I.S. Donnison, Influence of particle size on the analytical and chemical properties of two energy crops, Fuel 86 (2007) 60–72. https://doi.org/10.1016/J.FUEL.2006.06.022.

[140] C.A. Flores-Gómez, E.M. Escamilla Silva, C. Zhong, B.E. Dale, L. Da Costa Sousa, V. Balan, Conversion of lignocellulosic agave residues into liquid biofuels using an AFEX^{TM}-based biorefinery, Biotechnol Biofuels 11 (2018) 1–18. https://doi.org/10.1186/S13068-017-0995-6/FIGURES/6.

[141] H. Alizadeh, F. Teymouri, T.I. Gilbert, B.E. Dale, Pretreatment of switchgrass by ammonia fiber explosion (AFEX), Appl Biochem Biotechnol—Part A Enzyme Eng Biotechnol 124 (2005) 1133–1141. https://doi.org/10.1385/ABAB:124:1-3:1133.

[142] D. Wang, Y. Xin, H. Shi, P. Ai, L. Yu, X. Li, S. Chen, Closing ammonia loop in efficient biogas production: Recycling ammonia pretreatment of wheat straw, Biosyst Eng 180 (2019) 182–190. https://doi.org/10.1016/J.BIOSYSTEMSENG.2019.02.010.

[143] S. Sajad Hashemi, K. Karimi, A. Majid Karimi, Ethanolic ammonia pretreatment for efficient biogas production from sugarcane bagasse, Fuel 248 (2019) 196–204. https://doi.org/10.1016/J.FUEL.2019.03.080.

[144] M. del Mar Contreras-Gámez, Á. Galán-Martín, N. Seixas, A.M. da Costa Lopes, A. Silvestre, E. Castro, Deep eutectic solvents for improved biomass pretreatment: Current status and future prospective towards sustainable processes, Bioresour Technol 369 (2023) 128396. https://doi.org/10.1016/J.BIORTECH.2022.128396.

[145] Z. Chen, X. Bai, A. Lusi, H. Zhang, C. Wan, Insights into structural changes of lignin toward tailored properties during deep eutectic solvent pretreatment, ACS Sustain Chem Eng 8 (2020) 9783–9793. https://doi.org/10.1021/ACSSUSCHEMENG.0C01361/SUPPL_FILE/SC0C01361_SI_001.PDF.

[146] K.O. Olatunji, D.M. Madyira, Enhancing the biomethane yield of groundnut shells using deep eutectic solvents for sustainable energy production, Front Energy Res 12 (2024) 1346764. https://doi.org/10.3389/FENRG.2024.1346764.

[147] Y.L. Chen, X. Zhang, T.T. You, F. Xu, Deep eutectic solvents (DESs) for cellulose dissolution: A mini-review, Cellulose 26 (2019) 205–213. https://doi.org/10.1007/S10570-018-2130-7/METRICS.

[148] A. Procentese, F. Raganati, G. Olivieri, M.E. Russo, L. Rehmann, A. Marzocchella, Deep eutectic solvents pretreatment of agro-industrial food waste, Biotechnol Biofuels 11 (2018) 1–12. https://doi.org/10.1186/S13068-018-1034-Y/TABLES/4.

[149] A.D. Olugbemide, A. Oberlintner, U. Novak, B. Likozar, Lignocellulosic corn stover biomass pre-treatment by deep eutectic solvents (DES) for biomethane production process by bioresource anaerobic digestion, Sustainability 13 (2021) 10504. https://doi.org/10.3390/SU131910504.

[150] Y. Chen, T. Mu, Application of deep eutectic solvents in biomass pretreatment and conversion, Green Energy Environ 4 (2019) 95–115. https://doi.org/10.1016/J.GEE.2019.01.012.
[151] M. Sulaiman, F.A. Rabbani, T. Iqbal, M.A. Kazmi, S. Yasin, M.A. Mujtaba, M.A. Kalam, F. Almomani, Impact of eco-friendly chemical pretreatment on physicochemical and surface mechanical properties of sustainable lignocellulosic agricultural waste, Algal Res 71 (2023) 103051. https://doi.org/10.1016/J.ALGAL.2023.103051.
[152] W. Den, V.K. Sharma, M. Lee, G. Nadadur, R.S. Varma, Lignocellulosic biomass transformations via greener oxidative pretreatment processes: Access to energy and value-added chemicals, Front Chem 6 (2018) 337170. https://doi.org/10.3389/FCHEM.2018.00141.
[153] K.O. Olatunji, D.M. Madyira, Comparative analysis of the effects of five pretreatment methods on morphological and methane yield of groundnut shells, Waste Biomass Valorization 15 (2024) 469–486. https://doi.org/10.1007/S12649-023-02177-6/FIGURES/5.

5 Mechanical/physical pretreatments

5.1 INTRODUCTION

The ongoing demand for energy, driven by industrialization and human capital development, has increased the utilization of non-renewable fossil fuels, which poses a threat to the ecology [1]. Considering the projected population growth rate, global energy demand is expected to increase by 28% between 2015 and 2040 [2]. This projected increase is alarming considering the present oil reserves, greenhouse gas emissions, fuel costs, and pollution. Furthermore, concerns about climate change, global warming, national security, and instability have necessitated the need for alternative energy sources. Renewable energy, a clean energy source, has been identified as a potential alternative to fossil fuels. These renewable energy sources include biomass, wind, geothermal, solar, and hydroelectric [3], which can enhance energy efficiency and mitigate greenhouse gas emissions. Biomass is a natural, readily available, and carbon-neutral source of renewable energy for generating bioenergy and other valuable chemicals. This source provides a substantial feedstock for producing bioenergy, which helps minimize greenhouse gas emissions without compromising the food supply [3]. Lignocellulose biomass refers to wood or plant materials with primary constituents of cellulose, hemicellulose, and lignin, as shown in Figure 5.1. These constituents are strongly interwoven to form a complex arrangement, leading to the recalcitrant properties of the feedstock [4]. Bioenergy production from this lignocellulose biomass is not efficient due to its recalcitrant properties. Therefore, the application of pretreatment is required to break down the recalcitrant properties of this biomass and make it accessible during downstream bioprocessing. Pretreatment technique is classified into mechanical/physical, chemical, biological, thermal, and combined pretreatments [5]. Mechanical pretreatment involves the application of various treatments, including milling, grinding, high hydrostatic pressure (HHP), ultrasonic treatment, microwave treatment, and extrusion, to enhance the efficiency of bioenergy production. This technique is typically regarded as an initial step to reduce biomass size and enhance surface area before biofuel production or the application of other pretreatment methods. This chapter provides an overview of various mechanical pretreatment methods used in bioenergy production and the production of other value-added products. It also detailed the recent improvement in these methods and their challenges.

5.2 MILLING OR PARTICLE SIZE REDUCTION

Milling pretreatment is a popular mechanical technique used to reduce the particle size of lignocellulose biomass through shredding, grinding, chipping, and other

DOI: 10.1201/9781003633815-8

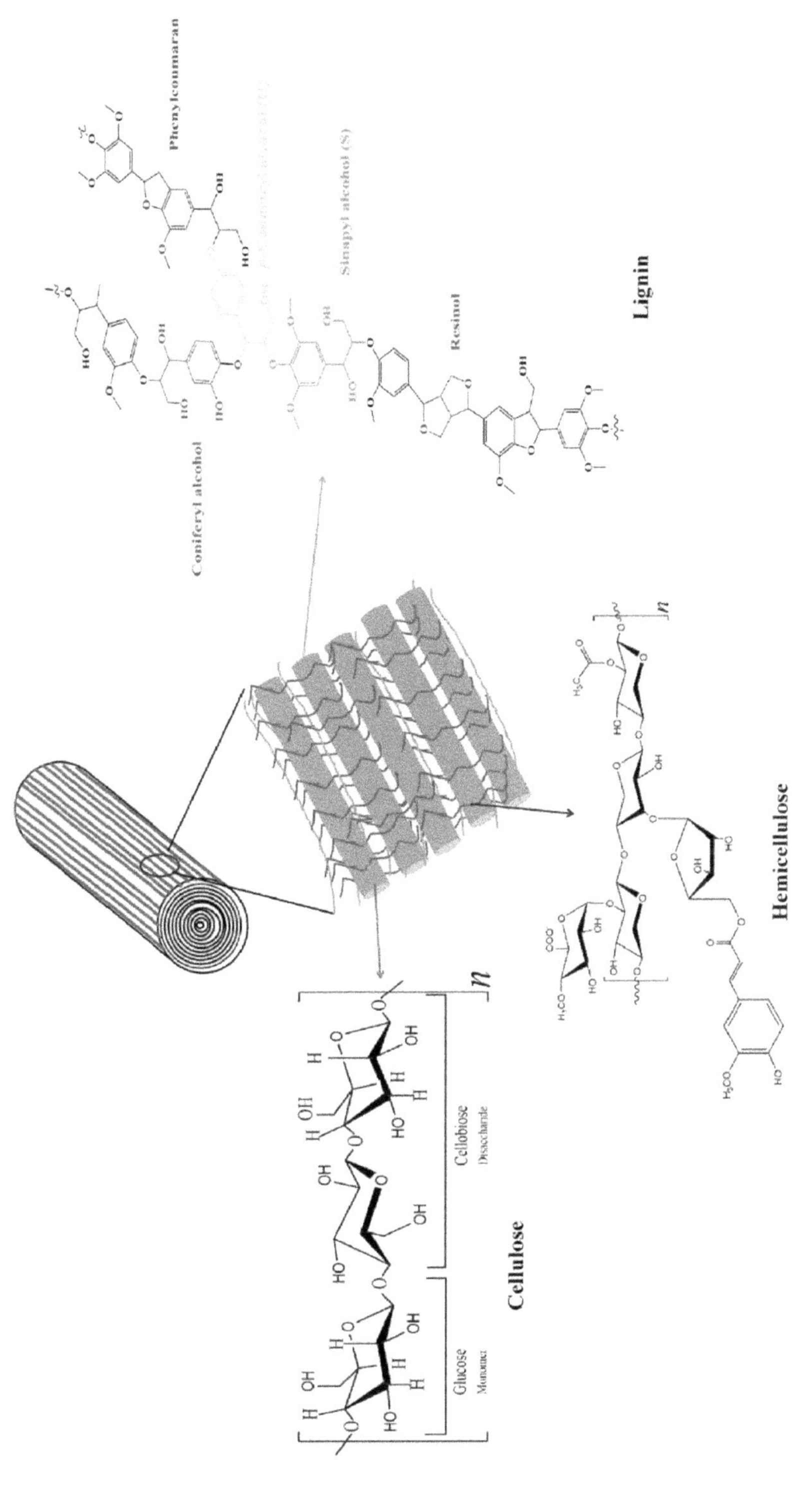

FIGURE 5.1 Structural arrangement of lignocellulose biomass [6].

methods, including those involving hammer, knife, ball, and attrition mills [7]. Biomass milling aims to reduce the degree of polymerization and cellulose crystallinity and improve surface area, hydrolysis efficiency, and mass transfer capacity [8]. Chipping and grinding are the most popular methods for size reduction in biomass pretreatment, with grinding being the most prominent. The word "grinding" is usually used interchangeably with other forms of size reduction in the biofuel industry. Biomass is primarily fractured into particulate forms with a mean diameter of less than 2 mm [9]. For chipping, the biomass can be reduced to between 10 and 30 mm, referred to as chips [9]. Hammer mill pretreatment can reduce the lignocellulose biomass to an average particle size of between 0.19 and 12.7 mm. The performance of a hammer mill is determined by various factors, including the aperture size of the screen, biomass type, and moisture content, among others [10]. This technique produces smaller biomass particles that are difficult to separate from one another due to further impaction. Knife milling can reduce the biomass particle to about 3.1 mm, with fewer commercial-scale applications due to the relatively poor throughput and high possibility of feedstock damage from pebbles and rocks [11].

Particle size reduction has been explored in various studies, yielding both remarkable successes and challenges. Arachis hypogea shells were reduced to 2-, 4-, and 6-mm particle sizes before anaerobic digestion. It was reported that 4-mm particle sizes released the optimum yield in terms of organic dry matter and on a fresh mass basis [12]. Pretreatment of rice straw and bagasse using particle size reduction shows that particle sizes below 1 mm released the highest biogas yield [13]. The effect of particle size reduction on biogas yield from municipal organic solid waste in Ouagadougou was investigated. Different particle sizes, ranging from 2000 to 63 μm, were experimented with, and the optimum biogas yield was recorded when a particle size of 250 μm was considered [14]. It can be observed that the optimum particle size varies for different feedstocks, depending on the morphological arrangement and functional groups of the biomass [15]. One of the major benefits of this technique is its ability to produce lower inhibitory compounds compared to some other techniques. Due to its ability to produce little to no furfural, 5-hydroxymethylfurfural, and phenols, it makes it the most suitable method for producing ethanol and methane. However, there are different challenges impeding the efficiency of this method. The energy required for particle size reduction is extremely high, significantly impacting the process's economic viability. Several factors, such as moisture content, initial particle size, mechanical properties, required biomass size, and feeding rate, must be considered when designing particle size reduction equipment to minimize energy requirements [16]. Another major limitation of this technique is its inability to remove lignin, which is a significant hindrance to bioenergy production from lignocellulose biomass [13]. Further studies are required to determine the optimal size requirement for most biomass and to identify the technique that can reduce energy demand for particle size reduction, making the technology more economical.

5.3 HIGH HYDROSTATIC PRESSURE (HHP)

HHP is a process that utilizes high pressure, in the order of MPa, transferred through an aqueous solution in a closed system. The technique has been in use for years within the food industry as non-thermal pasteurization, which involves the application of high pressure (100–600 MPa) without compromising the product's quality or the consumer's safety. The technology works on two fundamentals: (i) pressure is evenly distributed on the feedstock without considering the size and shape; and (ii) pressure favors all morphological reactions and alterations that include a reduction in volume [17]. The applied pressure significantly affects non-covalent bonds and influences the structures of macromolecules such as proteins and polysaccharides. The pressure also influences the hydrogen bonds but has no significant impact on the covalent bonds, which reduces the exposure time. When pressure is applied, it alters the arrangement of enzymes, affecting their performance and modifying their reaction mechanism, thereby disrupting the morphological structure of the biomass [18]. HHP pretreatment is also an efficient technique for protein degradation, which can modify the protein dynamics and degrade protein–protein bonds even without changes in temperature [19]. This method is a promising technology for biomass pretreatment for bioenergy, offering a specific set of properties that can be controlled based on the subsequent enzymatic hydrolysis. HHP pretreatment does not rely on mass/time, which is an advantage of the technology compared to thermal pretreatment.

The application of HHP pretreatment in bioenergy production has been investigated and reported to disrupt the microstructural arrangement of the biomass and enhance the downstream process during bioenergy production. It was observed that HHP pretreatment of coconut fibers, applied at a pressure range of 131–175 MPa, modifies the hydrogen bonds, ruptures the feedstock, and enhances the surface area [20]. Sugarcane bagasse pretreated with HHP using 400–800 MPa pressure combined with different concentrations of chemical solvents was noticed to improve the feedstock susceptibility and enzymatic hydrolysis and enhance the quantity of glucose released [21]. Tiny holes, cracks, and some fragments flaked off from the pretreated biomass were observed at 250 MPa, which is considered the optimum pressure. Pretreatment of corncob with HHP at 600 MPa combined with mild thermal treatment at 130 °C for 2 hrs was investigated. It was observed that the ferulic acid production was enhanced by 20% at 600 MPa for 15 min at 40 °C [22]. HHP is a promising pretreatment technique; however, it still faces several challenges that hinder its acceptability and application, particularly at the industrial scale. The need for special equipment that can withstand pressure is a significant challenge that must be addressed to make the process more economical. Literature on the impact of this technique on some readily available biomass for bioenergy production is still limited.

5.4 EXTRUSION

Extrusion is a mechanical pretreatment method where feedstock is forced through a die with a specified cross section. In this method, extruder screws (single or twin screws) are fixed in a tightly sealed barrel, which is used to break down the biomass. The rotation of the screws produces high shear forces between the barrel, screw, and biomass when the feedstock is introduced through the extruder [23]. It cuts, mixes, and heats to alter the microstructure of the biomass. It is a bright technique with high mechanical shear force to break the biomass arrangement, leading to the shortening and defibrillation of the biomass [24]. The effects of temperature and high-pressure during pretreatment modify the physical and chemical properties of feedstock, which improves the surface area and improves the enzymatic hydrolysis by making the biomass accessible to the enzymes [6]. Extrusion pretreatment occurs in three distinct reaction stages: conveying, reverse, and kneading. The biomass is automatically squeezed at the conveying stage before being transferred to the kneading stage, where it is mixed with the catalyst. It is then transferred back to the conveying stage, where the reactor maintains the required pressure for the technique. The biomass movement is continuous between the conveying and kneading stages during pretreatment [25]. The performance of extrusion pretreatment depends on the screw type, feedstock type, rotating screw speed, moisture, temperature, etc. [5]. Different studies have reported that extrusion pretreatment can improve the efficiency of the downstream bioenergy production process.

Vine-trimming shoots for biomethane production were pretreated using the extrusion technique with a twin-screw extruder, resulting in a 21% increase in biogas yield compared to the control sample. It was observed that the technique reduces the hemicellulose content by half and simultaneously enhances the quantity of soluble chemicals (lipids, proteins, carbohydrates, vitamins, and minerals) that are quickly converted by methanogenic microbes into methane, thereby increasing the gas released [26]. Extrusion pretreatment of rice straw was observed to improve the specific methane yield by 32.5% compared to milling pretreatment and 72.2% more than the untreated samples [27]. The use of twin-gear to pretreat rice straw enhances the amorphous region, leading to a crystallinity index increase of approximately 50%. Sharper peaks of cellulose were observed, which improved the glucan percentage to 63.16% and removed lignin up to 64.51% [28]. This technique has several merits, including high throughput, short exposure, continuous or semi-continuous process, post-processing cost (conditioning and washing) is not required, 50–80% solid loading rate, generates no effluent, and produces little or no inhibitors. The technique also features effective heat-mass transfer and temperature regulation, in addition to its ease of adaptability to various feedstocks [29,30]. However, the application of extrusion pretreatment is limited due to several challenges that require further investigation to enhance the technique's efficiency. There is a challenge of biomass flow capacity restriction that causes solid and liquid separation during pretreatment. Although the addition of a chemical can be used to overcome this, this could lead to the release of an inhibitory product. Extrusion at high pressure can also lead to the generation of an inhibitory compound [31]. The efficiency of this method can be enhanced by combining it with other pretreatment methods, which can be investigated in future studies.

5.5 GAMMA-RAY IRRADIATION

The gamma-ray irradiation pretreatment technique is a bright technique for lignocellulose biomass in bioenergy generation. It has been observed to break down complex biomass into an amorphous arrangement and enhance the cellulose conversion into fermentable sugars [32]. The gamma-ray irradiation pretreatment involves applying γ-rays from radioactive materials, such as cobalt-60 or cesium-137, to degrade the fiber lattice and activate the lignocellulose molecules, producing free radicals that subsequently depolymerize the biomass. These rays sterilize, inhibit, or support growth and modify the material's characteristics [33]. Ionizing radiation can easily penetrate biomass, degrading the lignin structure and altering the cellulose crystal areas. The resultant effect is facilitated by the generation of free radicals that degrade faster from the amorphous region when the radiation is stopped, while decay at a certain time from the crystalline region also results in further degradation of the biomass [34]. Gamma rays from radioisotopes can excite and charge the molecules, eventually leading to chain breakage and depolymerization or chain bond-linking reactions, which are crucial for cross-linking and influencing morphology [35]. It can generate some active particles in the primary components of the biomass cell walls, leading to a series of alterations in the biomass characteristics, like mechanical strength, molecular arrangement, dimensional stability, and hygroscopicity [36].

Wheat straw was pretreated using gamma-ray pretreatment with varying doses of Co-60 γ-radiation. It was observed that a significant impact was observed on the feedstock at 500 kGy, resulting in an improved glucose yield. The pretreated feedstock produced increased amounts of sugar (xylose and glucose), and no traces of glucuronic acid were detected [37]. Different doses of 50, 100, 150, 200, 250, and 300 Gy gamma-ray irradiation were used to pretreat *Brachypodium distachyon*. It was observed that the pretreatment had a negative impact on the feedstock, influencing its morphology, including cell size and biomass height, by increasing the tiller quantity and internode radius. It was observed from the scanning electron microscope image that the cell wall surface was reduced by 45% at 300 Gy compared to the untreated sample. The glucose released from samples pretreated at 250 and 300 Gy was improved by 121.5 and 139.4%, respectively, compared to the untreated feedstock [38]. When rapeseed straw was pretreated with gamma-ray irradiation at 1200 kGy, the morphological arrangement of the biomass was significantly physically and chemically altered. Pretreatment breaks down the carbohydrate bond and links between the feedstock, enhances the surface area, lowers the distribution range, and reduces the thermal stability [35]. An increase of 22% in biogas released when the sewage sludge cell was pretreated with gamma-ray pretreatment [39]. The main merits of this technique are that it requires mild temperature, short exposure time, does not require washing after pretreatment, and releases lower inhibitory compounds compared to chemical pretreatments [40]. The appropriate selection of process parameters, particularly γ-irradiation, can mitigate the harmful effects of various compounds produced by irradiation during downstream processes. Despite the merits of this technique, there are still some drawbacks that limit its application. The technique requires specific equipment, which limits the quantity of feedstock that can be treated at a time, presenting some safety challenges. The relationship between the gamma-ray dose and the degradation rate of the biomass has yet to be established. Likewise, the systematic investigation of the physicochemical characteristics of several biomass samples pretreated with gamma-ray irradiation has yet to be reported [41]. Exposure to high doses of irradiation in an oxygen-rich environment can produce inhibitory materials, such as 2-oxygluconic acid, hydrogen, CO_2, CO, carboxyl, and carboxylate compounds, which negatively impact yeast fermentation during biofuel production [40].

5.6 ULTRASOUND/SONICATION

Ultrasound is a sound wave with frequencies higher than the human hearing range (>20 kHz) and can be applied for biomass pretreatment. Biomass pretreatment using ultrasound typically occurs at a frequency of approximately 24 kHz and an operating power of 400 W. It is an advanced method of pretreating lignocellulose biomass that affects both chemical and physical characteristics of lignocellulose feedstocks. Ultrasonication is a mechanical technique that has been widely applied to pretreat biomass and can be combined with pH control to produce a combined physicochemical pretreatment method [42]. Cell degradation and lysis can occur through the combination of high process temperatures, pressures, and shear forces, which release highly reactive radicals during ultrasound pretreatment [43]. Ultrasound application to biomass produces acoustic cavitation, resulting in the generation of microbubbles. These microbubbles break down, releasing shockwaves that generate heat (up to 5000 °C), hydroxyl radicals (OH^-), and hydrogen (H^+), which leads to the disintegration of biomass by bursting the glycosidic links, as shown in Figure 5.2 [11]. This process is eco-friendly and effective pretreatment that breaks the strong arrangement of cellulose and hemicellulose, improving the porosity of the cellulose portion and resulting in the conversion into simple reducing sugars. The alteration strength of this technique reduces the lignin content, alters the functional groups, and enhances the crystallinity index while also improving the surface area and porosity of the feedstock [44]. The performance of this technique is determined by the type of

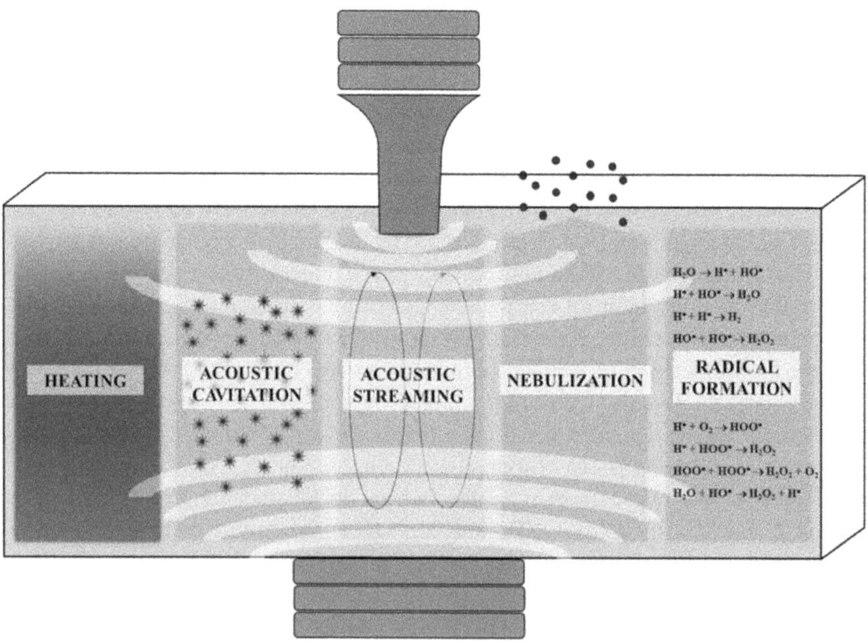

FIGURE 5.2 Molecular mechanisms of ultrasound biomass pretreatment [50].

biomass, sonication amplitude, and exposure time [8]. Additionally, the efficiency of the technique can be improved by combining it with other pretreatment methods, such as alkaline or acidic treatments.

Sorghum residue pretreated using ultrasound for 5 hrs at 50 °C and 40 kHz frequency was observed to increase the sugar released by 33.7% [45]. Bioenergy from algal biomass was improved by 7-fold when ultrasonic pretreatment was applied. It was also observed that longer exposure time favors the yield over a shorter treatment period [46]. Energy grass pretreated at 50% amplitude for 1 hr at 30 °C was observed to enhance the furfural released to approximately 72.4 mg/g [47]. When a similar ultrasonication pretreatment technique was applied to sugarcane bagasse, 785 lignin removal and 94.9% holocellulose recovery were recorded [48]. Ultrasound pretreatment of coconut coir, pistachio, and groundnut shells at 100 W ultrasound power at a feedstock loading rate of 0.5% (w/v) for 70 min was observed to improve the delignification from 80 to 100% [49]. This method has been investigated to improve bioenergy production at the laboratory scale; however, there is still limited literature on the techno-economic feasibility of the method for different biomasses at the industrial scale. Reactor design is another challenge facing the scalability of this technique. There is a need for the design of improved reactors that can reduce the energy required, which in turn will reduce operational expenses at the industrial scale [7].

5.7 ELECTRON BEAM (EB) IRRADIATION

Electron beam irradiation (EBI) was previously known for its application in sterilizing utensils before it was investigated as a potential technology for bio-restructuring biomass [51]. An EB is a stream of highly charged ionizing radiation generated from a linear electron accelerator [42]. This beam can be directed to biomass to alter the microstructural arrangement and disrupt the lignin, cellulose, and hemicellulose content. It transferred energy to the biomass atoms and disrupted the cell wall polymers, producing free radicals that move freely to break the cross-link structure of bond scission, reducing the polymerization degree, and decrystallization [52,53]. The irradiation radicals serve as the primary ionizing irradiation impacts and produce a higher level of disruption and polymerization through the possible breakdown of glycosidic linkages or cleavage in the biomass [54]. EBI irradiation of biomass requires a solvent, such as OH links from H_2O, to enhance the disruption of the biopolymeric arrangement. This pretreatment also influences the molar mass of the cellulose portion significantly [54]. EBI pretreatment can be used as a single pretreatment or combined with other pretreatment techniques. Beam current, acceleration voltage, and irradiation breadth are the primary parametric settings in the EB processing system, rather than the handling speed of the biomass on the conveyor, the required dose, and the process conditions of either an inert gas atmosphere or air [55].

The effect of EBI pretreatment on birch biomass was investigated at different doses prior to bioenergy production. It was established that the technique can degrade lignin and reduce the enzymatic resistance during biofuel production. Additionally, pretreatment significantly reduces the degree of polymerization, crystallinity, and cellulose molecular arrangement. It was also observed that the crystalline index of the sample decreased from 86.47 to 75.84% as the dose increased from 0 to 900 kGy, and depolymerization decreased from 2160 to 245. The degradation was reported as 9.09% when the pretreatment dose was increased to 1200 kGy. Total sugar yield from the feedstock was improved by 256.48% at a 1200 kGy dose compared to the untreated sample. An increase in dose was observed to enhance the oxidative degradation of 5-HMF and furfural, thereby minimizing the inhibitory effect [56]. Rice straw pretreated with EBI enhanced the digestion rate and conversion potential. It was also noticed that the process did not release significant inhibitory material [57]. Hemicellulose and cellulose contents of sugar maple were depolymerized at varying degrees, and phenolic compounds generated were increased when a 1000 kGy dose of EBI was applied during pretreatment [58]. Bamboo chips pretreated at low EBI between 0.5 and 50 kGy were observed to degrade cellulose significantly, while there was no significant influence on the hemicelluloses [59]. This indicates that treatment at a lower range is selective for cellulose, and applied power is a crucial parameter. The technique is clean, effective, and economical because temperature and pressure variations are not required, and little to no chemicals/solvents are needed, which is a major merit of the method. One of the limitations of this method is that the effect on lignin and hemicellulose is poor; therefore, there is a need to combine it with other pretreatment methods, such as steam explosion or chemical degradation of the hemicellulose and lignin. This can increase the cost of pretreatment and the overall production cost, and the possibility of releasing inhibitory compounds due to the combined use of other methods is high.

5.8 MICROWAVE IRRADIATION

Microwave irradiation pretreatment utilizes electromagnetic radiation with frequencies ranging from 0.3 to 300 GHz to pretreat lignocellulose biomass, as illustrated in Figure 5.3. It utilizes the magnetic and electric field components of microwaves to generate forces that rapidly change orientation at a rate of 2.4 x 10^9 times/sec. These low-frequency non-ionizing waves are produced by reversing the flow of a dipole, where the dipole is regarded as the dissociation of two opposite but equal charges, separated by a fixed distance [60]. The movement of polar molecules and ions during the process results in physical, biological, and chemical disruptions due

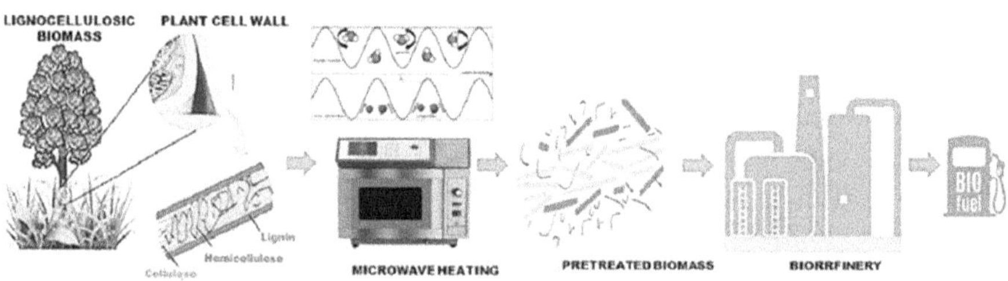

FIGURE 5.3 Microwave irradiation pretreatment [63].

to the extensive collisions and heat that occur. The technique can be explained in terms of both thermal and non-thermal influences. Thermal impacts lead to lignin and cell wall degradation through the rise in pressure and temperature. Whereas the non-thermal impact forces are produced from relaxation and polarization of dielectric materials under the control of an electromagnetic field [61]. This process causes an explosion within the feedstock and degrades the recalcitrant characteristics, and improves the rearrangement of the crystalline contents. Recently, this technique has gained attention in biorefineries due to its fast and evenly distributed heating, reduced energy activation requirements, rapid kinetic reaction, lower inhibitory generation, and higher bioenergy release [8]. The performance of microwave irradiation pretreatment is determined by several factors, including microwave power, the molecular dynamics of the dipole molecule, feedstock type, feedstock size, associated temperature, penetration depth, moisture content, ionic conduction of electrolytes, and the induction of magnetic materials that produce Joule heating, as well as exposure time [61,62]. The efficiency of this technique can also be improved by combining it with other pretreatment techniques, such as alkali, acid, oxidative, milling, and bacterial treatments.

The effects of microwave irradiation pretreatment have been experimentally investigated on different feedstocks and reported to improve bioenergy recovery. A glucan conversion rate of approximately 89% was observed when wheat and rye stillage was pretreated for 15 min at 300 W and 54 PSI during microwave irradiation pretreatment [64]. Maize distillery stillage for bioethanol production was pretreated using microwave-assisted pretreatment at 300 W for 15 min at 54 PSI. It was observed that high glucose levels of 104.4 ± 0.4 mg/g dry weight and an optimum enzymatic hydrolysis of 75.8 ± 0.9% were achieved. It was also reported that the treatment conditions reduce the amount of furfural and 5-hydroxymethylfurfural generation [65]. Pineapple peels were pretreated with microwave irradiation using a microwave power of 90–900 W, an exposure time of 5–20 min, and a loading rate of 100–160 g/l. It was observed that an optimum sugar release of 80.2% was achieved at 900 W for 9 min at a loading rate of 100 g/l, resulting in a glucose yield of 7.8% [66]. Brewery spent grains (BSG) were previously pretreated using the microwave hydrothermal technique for butanol production. The findings show that at 192.7 °C for 5.4 min, 64% of the hemicellulose sugar was recovered, and enzymatic hydrolysis produced a 70% glucose recovery, accompanied by the generation of 2.4 g/l of inhibitors. Enzymatic hydrolysis of the sugar solution fermented produced a butanol concentration and a total yield of 8.3 g/l and 46 kg/t BSG, respectively [67]. Microwave-alkali pretreatment of switchgrass was noticed to release 70–90% of sugars, and NaOH was identified as the most suitable alkali [68]. In a similar study, coastal Bermuda and switchgrass were subjected to a microwave-based alkali method using various alkali solutions. It was reported that coastal Bermuda grass released 59% xylose and 87% glucose, whereas switchgrass generated 63% xylose and 82% glucose [69]. This technology has been identified as an alternative method for the cost-efficient biorefinery of lignocellulose biomass due to its low energy and reagent requirements. The technique is easy and self-reliant. Nonetheless, there are some challenges limiting the application of microwave pretreatment. Microwave irradiation pretreatment is not suitable for all feedstocks, especially dry feedstocks, but it is mostly efficient in feedstocks with high moisture content. Another limitation is the inability to monitor the temperature within the microwave, which can lead to thermal destruction of the pretreated biomass due to the non-uniformity of the heating behavior. There is a need to design specific equipment that can be used to increase the cavity to improve the homogeneity of the microwave field. The tendency for microwave leakage poses a significant risk to human health (HH) and necessitates special safety protocols and precautions [60].

5.9 HIGH-PRESSURE HOMOGENIZATION (HPH)

This is a popular mechanical pretreatment method that is known for its ability to degrade cells and release intracellular bioproducts. It utilizes a homogenizer to homogenize the feedstock by forcing liquid through a specific valve using a pressure pump, thereby producing a uniform particle size distribution in the liquid [70]. This technology applies the fundamental principle whereby the applied pressure is distributed evenly across the feedstock, irrespective of its shape and size. The second principle entails that the pressure is favorable to all structural reactions and alterations, including volume reduction. The method can be referred to as HPH when the pressure is around 150–200 MPa or ultra-high-pressure homogenization (UHPH) when the pressure is around 350–400 MPa [71]. The pressure impacts the hydrogen bonds only, with no significant impact on the covalent bonds, thereby reducing the exposure time. It can alter the protein arrangement and impact the protein–protein bonds [19]. Applied pressure affects the morphological arrangement and enhances the activities of enzymes, altering their reaction mechanisms that break down the biomass's morphological structure [18]. This method is a promising approach for biomass pretreatment when a specific property is required, as pressure and exposure time can be controlled to achieve the desired degree of degradation.

Different studies have reported the use of HHP for the solid and liquid states of sludge biomass [71]. The application of HHP pretreatment to coconut fibers was observed to break the hydrogen bonds, rupture the feedstock, and produce a porous material when the pressure was between 131 and 175 MPa [72]. Biomass degradation with a high-pressure homogenizer was observed to reduce volatile solids and improve suspended solids with an increase in pressure and cycle [73]. The reduction of solids and liquefaction of volatile solids supported biogas yield enhancement. HHP pretreatment of Eucalyptus globulus kraft pulp at 300–400 MPa for 15–45 min was observed to increase the hydrolysis rate of xylan by xylanase by 5–10-fold [74]. HPH at 10 MPa was applied to different biomass sources, including catalpa sawdust, grass clipping, pine sawdust, and corn straw, to improve their digestibility. It was observed that pretreatment significantly disrupts the biomass by reducing the particle sizes, denaturing the structure, and modifying the crystallinity. The reducing sugar produced for subsequent biofuel production was also improved. Grass clipping pretreated was observed to release 229.42 mg/g optimum sugar when the sample underwent enzymatic hydrolysis for 48 hrs. Enzymatic hydrolysis of pretreated corn straw was enhanced by 68.37%, but pine sawdust and catalpa sawdust generated very low reducing sugar [75]. Reports on the influence of HHP pretreatment on the physical and chemical characteristics of biomass are very limited in the literature. This limits the application of this technique because information about impacts on physical and chemical properties is crucial to the adoption of pretreatment techniques. The cost of pretreatment using this technique is slightly higher due to the requirement for higher energy to generate the pressure and the need for specialized equipment. Further research is recommended in this area for more information on the impacts on physicochemical properties and the production of equipment with lower energy requirements.

5.10 PULSE-ELECTRIC FIELD

Pulse-electric field pretreatment is a technique that degrades biomass using simple equipment with two electrodes, where biomass is introduced to a non-thermal voltage pulse for a brief exposure period (nanoseconds to milliseconds) under a high pulse amplitude of 0.1–100 kV/cm [76]. Biomass exposure to this technique causes degradation and structural alterations in the cell membrane, leading to the loss of semi-permeability and enhancing the accessibility of the hydrolyzing enzyme to lignin and polysaccharides [77]. This technique can induce osmotic stabilization on lignocellulose and algal biomass. It can have varying influences on cell degradation, temperature elevation, mechanical fragmentation of the feedstock, and other properties of pretreated biomass. It is advisable to exempt the electric arc phenomena when using this method. Generally, this technique is used in a relatively homogeneous electric field (between plate electrodes) as illustrated in Figure 5.4 [76]. The effect of

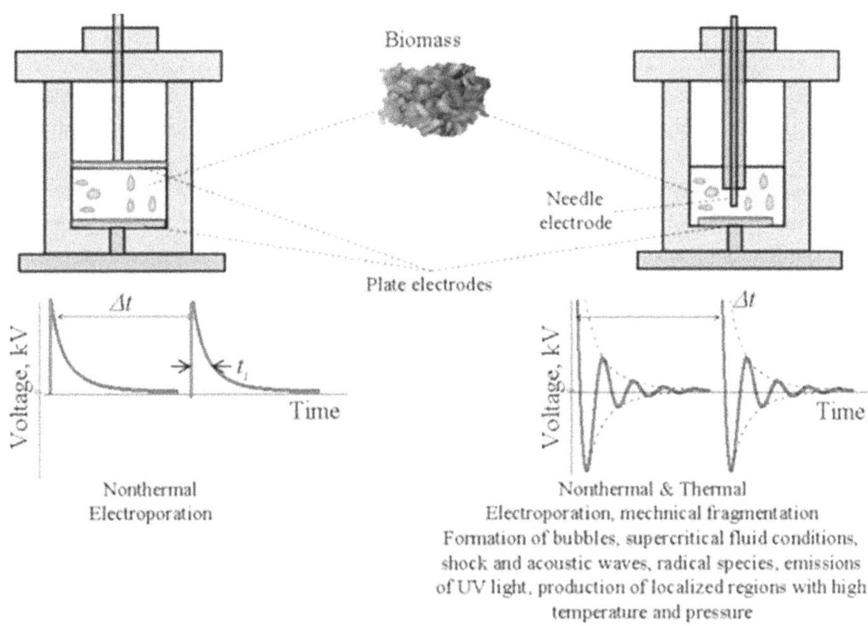

FIGURE 5.4 Pulse-electric field pretreatment process [76].

this pretreatment technique on biomass is primarily due to the loss of the membrane barrier, a phenomenon known as electroporation [78]. The electrical potential (u_m) applied to the biomass defines the level of electroporation. Electroporation occurs at voltages greater than 0.5–1.5 V, and the cell radius is directly proportional to the voltage value. This indicates that bigger cells are degraded before smaller cells during pretreatment. Properties of the biomass, such as electrical conductivity and cell size, and pulse parameters, including pulse amplitude (or electric field strength), exposure time, shape, pulse interval, and number of repeats, determine the degree of electroporation. The most used pulse shapes for biomass pretreatment are near-rectangular or exponential decay, which can be monopolar or bipolar. Findings have shown that crucial disruption to the biomass can be achieved when electric field strengths (E) are between 500 and 1000 V/cm and exposure times (t_{PEF}) are between 10^{-4} and 10^{-1} sec [79]. A pulse-electric field is usually utilized under non-thermal conditions, with no significant increase in temperature. However, it should be noted that the process may exhibit thermal synergistic influences and a tendency toward drying, mild thermal, cold diffusion, and freezing treatments [80].

A pulse-electric field was applied to the wood chip and switchgrass using an electric field capacity of 10 kV/cm with 2000 pulses, and the treatment enhanced the porosity of the samples, which shows its possibility to enhance the cellulose hydrolysis [81]. 73.8% carbohydrate extraction, which was a 10% increase compared to the untreated sample, was observed when a pulse-electric field with 20 kV/cm and 100 kJ/kg was applied to A. platensis microalgae at 35 °C [82]. Biomethane yield from pig slurry was improved by 80% when pretreated with a pulse-electric field, while biomethane yield from waste-activated sludge was enhanced by 100% under the same treatment conditions [83]. The merits of pulse-electric field include a low energy requirement and the avoidance of unwanted modifications in biomass, which are commonly observed in other pretreatment methods, such as chemical and thermal techniques. The technique can also be an alternative to other techniques with high energy or temperature requirements [84]. Pulse-electric field pretreatment aligns with modern biomass pretreatment methods. This technique is in its infancy; therefore, further investigation is needed to enhance its efficiency in biomass pretreatment. Sustainable biorefinery applications require large-scale units to achieve high-quality flow rates, which may necessitate the use of multiple pulse generators, potentially leading to increased production costs.

5.11 FREEZING PRETREATMENT

Freezing pretreatment is a technique that pretreats biomass by inducing volumetric changes in water during the freezing process. It is a relatively new and promising technique that has been identified as a means to enhance the efficiency of the downstream bioprocess [85]. In this technique, biomass undergoes different cycles of very low temperature variations. It involves mixing biomass with a certain volume of water and placing it in a freezer until the water freezes. Water penetrates the biomass, and as the water volume expands during freezing, it causes degradation of the cell walls. The process leads to the production of intracellular ice crystals, which result in swollen cells and the cellular assemblage being degraded, thereby breaking the overall biomass arrangement [86]. This opened the cellulose content and improved the available surface area for further bioprocessing. The technique does not require the addition of chemicals or catalysts, making it an economically inexpensive and environmentally benign technique. This technique can be used to pretreat biomass in cold regions of the world when wet biomass is placed in open space, allowing the sample to freeze and expand due to temperature variations in the open space. In this situation, no energy is required for the pretreatment process.

Milled barley straw was frozen at −18 °C for a specified number of times and thawed at about 22 °C (room temperature) to pretreat the sample. To understand the influence of freezing on biomass cycles, the process was examined at different cycles and compared with the untreated sample. It was observed that the optimum hydrolysis efficiency of 19.42% was achieved when the sample was cycled four times, while the untreated sample produced 10.28%. The sample fermentation after pretreatment was observed to increase to approximately 88.80 g/kg of the barley straw [86]. Freezing pretreatment of rice straw was observed to improve the enzymatic hydrolysis of rice straw, yielding 417.27 g/kg of sugar when treated with 150 U cellulase and 138.77 g/kg when treated with 100 U xylanase for 48 hrs [87]. It was concluded that freezing pretreatment can be used as the initial stage in a combined pretreatment process. Freezing pretreatment requires a longer exposure time and high energy to freeze the biomass, making the technique inefficient. More studies are still required to make the method acceptable, especially on an industrial scale.

5.12 ENVIRONMENTAL AND SOCIAL IMPACTS

Mechanical pretreatment is the most commonly used pretreatment method, especially particle size reduction, which is often required for most lignocellulose biomass before biorefinery or treatment with another method. The primary target of mechanical pretreatment is to reduce particle size, depolymerize, and decrease crystallinity, thereby improving availability and accessibility during the downstream process. This technique is environmentally friendly, as it releases few harmful materials into the environment, since there is no need for washing, and minimal chemicals are required [70]. Life cycle assessment (LCA) can be used to analyze the environmental impact of mechanical pretreatment in the biorefinery according to the Tool for the Reduction and Assessment of Chemical and other Environmental Impacts (TRACI) method [88]. The environmental impact assessment of mechanical pretreatment, as defined by the TRACI technique, evaluates the effect through various parameters, including HH, particulate matter (PM2.5 eq/kg substance), acidification (kg SO_2 eq/kg substance), eutrophication potential, and energy required [89]. The acidification potential of mechanical pretreatment is lower compared to other techniques, such as chemical pretreatment. This can be increased when the method is combined with other techniques, such as microwave-assisted alkali pretreatment. When HH and eutrophication factors are considered, mechanical pretreatment is a more effective option compared to other methods. The energy required for mechanical pretreatment is high, based on the available data, compared to most other methods. This level of energy required has made mechanical pretreatment one of the most expensive techniques in biorefinery in terms of operating cost. This challenge can be overcome when multi-biorefinery or co-product recovery is considered in a biorefinery setup [90]. There is a limited literature on the techno-economic analysis of mechanical pretreatment of lignocellulosic biomass, which makes comparison challenging.

5.13 CONCLUSION

Mechanical pretreatment of lignocellulose biomass in a biorefinery aims to break down the recalcitrant properties of the biomass. This technique typically reduces the size of the biomass but does not always separate the lignin, hemicellulose, and cellulose to a significant degree, as they are strongly intertwined in the biomass's morphological arrangement. However, the mechanical disruption of the biomass was observed to enhance the efficiency of further treatments because of its ability to improve the contact surface area of the biomass. It was observed that different mechanical pretreatment techniques have varying impacts on the same biomass, and they do not produce the same level of improvement in different biorefinery processes. Studies have shown that mechanical pretreatments generally enhance bioenergy release when the appropriate conditions are selected; however, their inability to degrade lignin is a major limitation to the bioavailability of carbohydrates for biorefinery. The high energy required and the cost of energy, which make the technique exorbitant and unsustainable, are major limitations to mechanical pretreatment techniques. Therefore, further research and development are necessary to reduce energy consumption, enhance the efficiency of mechanical pretreatment, and improve the economic viability of biorefinery processes.

REFERENCES

[1] Q. Hassan, P. Viktor, T.J. Al-Musawi, B. Mahmood Ali, S. Algburi, H.M. Alzoubi, A. Khudhair Al-Jiboory, A. Zuhair Sameen, H.M. Salman, M. Jaszczur, The renewable energy role in the global energy Transformations, Renew Energy Focus 48 (2024) 100545. https://doi.org/10.1016/J.REF.2024.100545.

[2] World Energy Outlook 2024—Analysis—IEA (n.d.). https://www.iea.org/reports/world-energy-outlook-2024 (accessed July 30, 2025).

[3] B.R. Prasad, R.K. Padhi, G. Ghosh, A review on key pretreatment approaches for lignocellulosic biomass to produce biofuel and value-added products, Int J Environ Sci Technol 20 (2022) 6929–6944. https://doi.org/10.1007/S13762-022-04252-2.

[4] K.O. Olatunji, D.M. Madyira, Comparative analysis of the effects of five pretreatment methods on morphological and methane yield of groundnut shells, Waste Biomass Valoriz 15 (2024) 469–486. https://doi.org/10.1007/S12649-023-02177-6/FIGURES/5.

[5] K.O. Olatunji, N.A. Ahmed, O. Ogunkunle, Optimization of biogas yield from lignocellulosic materials with different pretreatment methods: A review, Biotechnol Biofuels 14 (2021) 1–34. https://doi.org/10.1186/S13068-021-02012-X.

[6] B. Kumar, N. Bhardwaj, K. Agrawal, V. Chaturvedi, P. Verma, Current perspective on pretreatment technologies using lignocellulosic biomass: An emerging biorefinery concept, Fuel Process Technol 199 (2020) 106244. https://doi.org/10.1016/J.FUPROC.2019.106244.

[7] M. Jędrzejczyk, E. Soszka, M. Czapnik, A.M. Ruppert, J. Grams, Physical and chemical pretreatment of lignocellulosic biomass, Second and Third Generation of Feedstocks: The Evolution of Biofuels (2019) 143–196. https://doi.org/10.1016/B978-0-12-815162-4.00006-9.

[8] A.R. Mankar, A. Pandey, A. Modak, K.K. Pant, Pretreatment of lignocellulosic biomass: A review on recent advances, Bioresour Technol 334 (2021) 125235. https://doi.org/10.1016/J.BIORTECH.2021.125235.
[9] Y. Sun, J. Cheng, Hydrolysis of lignocellulosic materials for ethanol production: A review, Bioresour Technol 83 (2002) 1–11. https://doi.org/10.1016/S0960-8524(01)00212-7.
[10] O. Oyedeji, P. Gitman, J. Qu, E. Webb, Understanding the impact of lignocellulosic biomass variability on the size reduction process: A review, ACS Sustain Chem Eng 8 (2020) 2327–2343. https://doi.org/10.1021/ACSSUSCHEMENG.9B06698/ASSET/IMAGES/MEDIUM/SC9B06698_0005.GIF.
[11] A.V.S.L. Sai Bharadwaj, S. Dev, J. Zhuang, Y. Wang, C.G. Yoo, B.H. Jeon, S. Aggarwal, S.H. Park, T.H. Kim, Review of chemical pretreatment of lignocellulosic biomass using low-liquid and low-chemical catalysts for effective bioconversion, Bioresour Technol 368 (2023) 128339. https://doi.org/10.1016/J.BIORTECH.2022.128339.
[12] S.O. Jekayinfa, A.O. Adebayo, O.O. Oniya, K.O. Olatunji, Comparative analysis of biogas and methane yields from different sizes of groundnut shell in a batch reactor at mesophilic temperature, J Energy Res Rev 5 (2020) 34–44. https://doi.org/10.9734/jenrr/2020/v5i130140.
[13] M. Kaur, Effect of particle size on enhancement of biogas production from crop residue, Mater Today Proc 57 (2022) 1950–1954. https://doi.org/10.1016/J.MATPR.2022.03.292.
[14] M. Nikiema, N. Barsan, A.S. Ouili, E. Mosnegutu, K.M. Somda, Y. Maiga, C.C.O. Tidiane, C.A.T. Ouattara, V. Nedeff, A.S. Ouattara, Determination of particle size for optimum biogas production from ouagadougou municipal organic solid waste, Sustainability 16 (2024) 9792. https://doi.org/10.3390/SU16229792.
[15] D.M. Madyira, K.O. Olatunji, Investigating the influence of different pretreatment methods on the morphological structure of Arachis hypogea shells, Mater Today Proc 105 (2024) 72–77. https://doi.org/10.1016/J.MATPR.2023.07.289.
[16] F.R. Amin, H. Khalid, H. Zhang, S. Rahman, R. Zhang, G. Liu, C. Chen, Pretreatment methods of lignocellulosic biomass for anaerobic digestion, AMB Express 7 (2017) 72. https://doi.org/10.1186/s13568-017-0375-4.
[17] D. Kim, G.D. Han, High hydrostatic pressure treatment combined with enzymes increases the extractability and bioactivity of fermented rice bran, Innov Food Sci Emerg Technol 16 (2012) 191–197. https://doi.org/10.1016/J.IFSET.2012.05.014.
[18] M.J. Eisenmenger, J.I. Reyes-De-Corcuera, High pressure enhancement of enzymes: A review, Enzyme Microb Technol 45 (2009) 331–347. https://doi.org/10.1016/j.enzmictec.2009.08.001.
[19] E. Ozturk, M.H. Oztop, H. Alpas, Use of high hydrostatic pressure (HHP) for increasing the product yield of lignocellulosic biomass hydrolysis: A study for peanut hull and microcrystalline cellulose, LWT 147 (2021) 111556. https://doi.org/10.1016/J.LWT.2021.111556.
[20] E.D. Albuquerque, F.A.G. Torres, A.A.R. Fernandes, P.M.B. Fernandes, Combined effects of high hydrostatic pressure and specific fungal cellulase improve coconut husk hydrolysis, Process Biochem 51 (2016) 1767–1775. https://doi.org/10.1016/J.PROCBIO.2016.07.010.
[21] J.F. Castañón-Rodríguez, B. Torrestiana-Sánchez, M. Montero-Lagunes, J. Portilla-Arias, J.A.R. De León, M.G. Aguilar-Uscanga, Using high pressure processing (HPP) to pretreat sugarcane bagasse, Carbohydr Polym 98 (2013) 1018–1024. https://doi.org/10.1016/j.carbpol.2013.06.068.
[22] N. Pérez-Rodríguez, A. Torrado Agrasar, J.M. Domínguez, High hydrostatic pressure as pretreatment and adjuvant for the enzymatic release of ferulic acid from corn cob, Process Biochem 58 (2017) 204–210. https://doi.org/10.1016/J.PROCBIO.2017.04.011.

[23] A. Duque, P. Manzanares, M. Ballesteros, Extrusion as a pretreatment for lignocellulosic biomass: Fundamentals and applications, Renew Energy 114 (2017) 1427–1441. https://doi.org/10.1016/J.RENENE.2017.06.050.
[24] B. Basak, S. Saha, P.K. Chatterjee, A. Ganguly, S. Woong Chang, B.H. Jeon, Pretreatment of polysaccharidic wastes with cellulolytic Aspergillus fumigatus for enhanced production of biohythane in a dual-stage process, Bioresour Technol 299 (2020) 122592. https://doi.org/10.1016/J.BIORTECH.2019.122592.
[25] Y.L. Cha, J. Yang, S. Il Seo, G.H. An, Y.H. Moon, G.D. You, J.E. Lee, J.W. Ahn, K.B. Lee, Alkaline twin-screw extrusion pretreatment of Miscanthus with recycled black liquor at the pilot scale, Fuel 164 (2016) 322–328. https://doi.org/10.1016/J.FUEL.2015.10.006.
[26] N. Pérez-Rodríguez, D. García-Bernet, J.M. Domínguez, Extrusion and enzymatic hydrolysis as pretreatments on corn cob for biogas production, Renew Energy 107 (2017) 597–603. https://doi.org/10.1016/J.RENENE.2017.02.030.
[27] X. Chen, Y.L. Zhang, Y. Gu, Z. Liu, Z. Shen, H. Chu, X. Zhou, Enhancing methane production from rice straw by extrusion pretreatment, Appl Energy 122 (2014) 34–41. https://doi.org/10.1016/J.APENERGY.2014.01.076.
[28] M.A. Ahmed, M.S.U. Rehman, R. Terán-Hilares, S. Khalid, J.I. Han, Optimization of twin gear-based pretreatment of rice straw for bioethanol production, Energy Convers Manag 141 (2017) 120–125. https://doi.org/10.1016/J.ENCONMAN.2016.06.022.
[29] K. Muthukumarappan, C. Karunanithy, Extrusion processing: Opportunities and challenges toward biofuel, Biomass Fractionation Technologies for a Lignocellulosic Feedstock Based Biorefinery (2016) 57–79. https://doi.org/10.1016/B978-0-12-802323-5.00003-7.
[30] Y. Zheng, J. Zhao, F. Xu, Y. Li, Pretreatment of lignocellulosic biomass for enhanced biogas production, Prog Energy Combust Sci 42 (2014) 35–53. https://doi.org/10.1016/J.PECS.2014.01.001.
[31] D. Konan, E. Koffi, A. Ndao, E.C. Peterson, D. Rodrigue, K. Adjallé, An overview of extrusion as a pretreatment method of lignocellulosic biomass, Energies 15 (2022) 3002. https://doi.org/10.3390/EN15093002.
[32] J. Chen, L. Wang, X. Su, K. Wang, X. Wu, L. Chen, X. Xiong, H. Zhou, Y. Liu, Structure, morphology, thermostability and irradiation-mediated degradation fractions of hemicellulose treated with Γ-irradiation, Waste Biomass Valo 7 (2016) 1415–1425. https://doi.org/10.1007/S12649-016-9489-1/TABLES/2.
[33] S. Kim, J.O. Jeong, S. Lee, J.S. Park, H.J. Gwon, S.I. Jeong, J.G. Hardy, Y.M. Lim, J.Y. Lee, Effective gamma-ray sterilization and characterization of conductive polypyrrole biomaterials, Sci Rep8 (2018) 1–10. https://doi.org/10.1038/s41598-018-22066-6.
[34] S. Hyun Hong, J. Taek Lee, S. Lee, S. Gon Wi, E. Ju Cho, S. Singh, S. Sik Lee, B. Yeoup Chung, Improved enzymatic hydrolysis of wheat straw by combined use of gamma ray and dilute acid for bioethanol production, Radiat Phys Chem 94 (2014) 231–235. https://doi.org/10.1016/J.RADPHYSCHEM.2013.05.056.
[35] C.Y. Zhang, X.J. Su, X.Y. Xiong, Q.L. Hu, S. Amartey, X.H. Tan, W. Qin, 60Co-γ radiation-induced changes in the physical and chemical properties of rapeseed straw, Biomass Bioenergy 85 (2016) 207–214. https://doi.org/10.1016/J.BIOMBIOE.2015.11.022.
[36] A. Kalawate, S. Mehetre, Isolation and characterization of mold fungi and insects infecting sawmill wood, and their inhibition by gamma radiation, Radiat Phys Chem 117 (2015) 191–197. https://doi.org/10.1016/J.RADPHYSCHEM.2015.08.016.
[37] K. qin Wang, X. yao Xiong, J. ping Chen, L. Chen, X. Su, Y. Liu, Comparison of gamma irradiation and steam explosion pretreatment for ethanol production from agricultural residues, Biomass Bioenergy 46 (2012) 301–308. https://doi.org/10.1016/J.BIOMBIOE.2012.08.013.

[38] J.Y. Kim, C.S. Na, D.S. Kim, J.B. Kim, Y.W. Seo, The effect of chronic gamma ray irradiation on lignocellulose of Brachypodium distachyon, Cellulose 22 (2015) 2419–2430. https://doi.org/10.1007/S10570-015-0687-Y/FIGURES/4.
[39] S. Fang, L. Ping, Z. Yang, J. Mao, A review of different pretreatment techniques for enhancing biogas production, in: ICMREE2011—Proceedings 2011 International Conference on Materials for Renewable Energy and Environment (2011) 263–266. https://doi.org/10.1109/ICMREE.2011.5930810.
[40] C. Tissot, S. Grdanovska, A. Barkatt, J. Silverman, M. Al-Sheikhly, On the mechanisms of the radiation-induced degradation of cellulosic substances, Radiat Phys Chem 84 (2013) 185–190. https://doi.org/10.1016/J.RADPHYSCHEM.2012.06.020.
[41] D. Adliene, Radiation interaction with condensed matter, Applications of Ionizing Radiation in Materials Processing 1 (2017) 33–54.
[42] S.S. Hassan, G.A. Williams, A.K. Jaiswal, Emerging technologies for the pretreatment of lignocellulosic biomass, Bioresour Technol 262 (2018) 310–318. https://doi.org/10.1016/J.BIORTECH.2018.04.099.
[43] J. Byun, Y.L. Cha, S.M. Park, K.S. Kim, J.E. Lee, Y.G. Kang, Lignocellulose pretreatment combining continuous alkaline single-screw extrusion and ultrasonication to enhance biosugar production, Energies 13 (2020) 5636. https://doi.org/10.3390/EN13215636.
[44] N. Akhtar, D. Goyal, A. Goyal, Characterization of microwave-alkali-acid pre-treated rice straw for optimization of ethanol production via simultaneous saccharification and fermentation (SSF), Energy Convers Manag 141 (2017) 133–144. https://doi.org/10.1016/J.ENCONMAN.2016.06.081.
[45] J. Zhang, H. Zhou, D. Liu, X. Zhao, Pretreatment of lignocellulosic biomass for efficient enzymatic saccharification of cellulose, Lignocellulosic Biomass to Liquid Biofuels (2020) 17–65. https://doi.org/10.1016/B978-0-12-815936-1.00002-2.
[46] J.A. Choi, J.H. Hwang, B.A. Dempsey, R.A.I. Abou-Shanab, B. Min, H. Song, D.S. Lee, J.R. Kim, Y. Cho, S. Hong, B.H. Jeon, Enhancement of fermentative bioenergy (ethanol/hydrogen) production using ultrasonication of Scenedesmus obliquusYSW15 cultivated in swine wastewater effluent, Energy Environ Sci 4 (2011) 3513–3520. https://doi.org/10.1039/C1EE01068A.
[47] C.A. Bizzi, D. Santos, T.C. Sieben, G.V. Motta, P.A. Mello, E.M.M. Flores, Furfural production from lignocellulosic biomass by ultrasound-assisted acid hydrolysis, Ultrason Sonochem 51 (2019) 332–339. https://doi.org/10.1016/J.ULTSONCH.2018.09.011.
[48] G. Ramadoss, K. Muthukumar, Mechanistic study on ultrasound assisted pretreatment of sugarcane bagasse using metal salt with hydrogen peroxide for bioethanol production, Ultrason Sonochem 28 (2016) 207–217. https://doi.org/10.1016/J.ULTSONCH.2015.07.006.
[49] P.B. Subhedar, P. Ray, P.R. Gogate, Intensification of delignification and subsequent hydrolysis for the fermentable sugar production from lignocellulosic biomass using ultrasonic irradiation, Ultrason Sonochem 40 (2018) 140–150. https://doi.org/10.1016/J.ULTSONCH.2017.01.030.
[50] E.M.M. Flores, G. Cravotto, C.A. Bizzi, D. Santos, G.D. Iop, Ultrasound-assisted biomass valorization to industrial interesting products: State-of-the-art, perspectives and challenges, Ultrason Sonochem 72 (2021) 105455. https://doi.org/10.1016/J.ULTSONCH.2020.105455.
[51] A. Azizan, N.A.A. Jusri, M.F.A. Rahman, I.S. Azmi, N. Ibrahim, R.M. Salleh, Combining water-soaking, ionic liquid, and electron beam irradiation pretreatments for enzymatic hydrolysis of cellulose, Cellulose (2025) 1–16. https://doi.org/10.1007/S10570-025-06681-X/FIGURES/8.
[52] S. Sundar, N.S. Bergey, L. Salamanca-Cardona, A. Stipanovic, M. Driscoll, Electron beam pretreatment of switchgrass to enhance enzymatic hydrolysis to produce sugars for

biofuels, Carbohydr Polym 100 (2014) 195–201. https://doi.org/10.1016/J.CARBPOL.2013.04.103.
[53] X. Ma, X. Zheng, M. Zhang, X. Yang, L. Chen, L. Huang, S. Cao, Electron beam irradiation of bamboo chips: Degradation of cellulose and hemicelluloses, Cellulose 21 (2014) 3865–3870. https://doi.org/10.1007/S10570-014-0402-4/FIGURES/2.
[54] O. Sarosi, I. Sulaeva, E. Fitz, I. Sumerskii, M. Bacher, A. Potthast, Lignin resists high-intensity electron beam irradiation, Biomacromolecules 22 (2021) 4365–4372. https://doi.org/10.1021/ACS.BIOMAC.1C00926/ASSET/IMAGES/LARGE/BM1C00926_0005.JPEG.
[55] S.A. Hashim, K.A. Bakar, M.N. Othman, Electron beam accelerator: A new tool for environmental preservation in Malaysia, AIPC 1482 (2012) 210–213. https://doi.org/10.1063/1.4757467.
[56] X. Fei, W. Jia, J. Wang, T. Chen, Y. Ling, Study on enzymatic hydrolysis efficiency and physicochemical properties of cellulose and lignocellulose after pretreatment with electron beam irradiation, Int J Biol Macromol 145 (2020) 733–739. https://doi.org/10.1016/J.IJBIOMAC.2019.12.232.
[57] J.S. Bak, Process evaluation of electron beam irradiation-based biodegradation relevant to lignocellulose bioconversion, Springerplus 3 (2014) 1–6. https://doi.org/10.1186/2193-1801-3-487/TABLES/3.
[58] O.D. Mante, T.E. Amidon, A. Stipanovic, S.P. Babu, Integration of biomass pretreatment with fast pyrolysis: An evaluation of electron beam (EB) irradiation and hot-water extraction (HWE), J Anal Appl Pyrolysis 110 (2014) 44–54. https://doi.org/10.1016/j.jaap.2014.08.004.
[59] M.A. Kassim, H.P.S.A. Khalil, N.A. Serri, M.H.M. Kassim, M.I. Syakir, N.A.S. Aprila, R. Dungani, Irradiation pretreatment of tropical biomass and biofiber for biofuel production, Radiation Effects in Materials (2016). https://doi.org/10.5772/62728.
[60] N.R. Syed, B. Zhang, S. Mwenya, A.S. Aldeen, A systematic review on biomass treatment using microwave-assisted pyrolysis under PRISMA guidelines, Molecules 28 (2023) 5551. https://doi.org/10.3390/MOLECULES28145551.
[61] S. Tsubaki, J.I. Azuma, T. Yoshimura, M.M. Maitani, E. Suzuki, S. Fujii, Y. Wada, Microwave-induced biomass fractionation, Biomass Fractionation Technologies for a Lignocellulosic Feedstock Based Biorefinery (2016) 103–126. https://doi.org/10.1016/B978-0-12-802323-5.00005-0.
[62] A. Sidana, S.K. Yadav, Recent developments in lignocellulosic biomass pretreatment with a focus on eco-friendly, non-conventional methods, J Clean Prod 335 (2022) 130286. https://doi.org/10.1016/J.JCLEPRO.2021.130286.
[63] S. Ethaib, Microwave-assisted pretreatment for lignocellulosic biomass energy conversion path, Bioresour Technol Rep 28 (2024) 102006. https://doi.org/10.1016/J.BITEB.2024.102006.
[64] D. Mikulski, G. Kłosowski, Microwave-assisted dilute acid pretreatment in bioethanol production from wheat and rye stillages, Biomass Bioenergy 136 (2020) 105528. https://doi.org/10.1016/J.BIOMBIOE.2020.105528.
[65] D. Mikulski, G. Kłosowski, A. Menka, B. Koim-Puchowska, Microwave-assisted pretreatment of maize distillery stillage with the use of dilute sulfuric acid in the production of cellulosic ethanol, Bioresour Technol 278 (2019) 318–328. https://doi.org/10.1016/J.BIORTECH.2019.01.068.
[66] S. Chongkhong, C. Tongurai, Optimization of soluble sugar production from pineapple peel by microwave-assisted water pretreatment, Songklanakarin J Sci Technol 41 (2019) 237–245. https://doi.org/10.14456/sjst-psu.2019.29.

[67] J.C. López-Linares, M.T. García-Cubero, S. Lucas, G. González-Benito, M. Coca, Microwave assisted hydrothermal as greener pretreatment of brewer's spent grains for biobutanol production, Chem Eng J 368 (2019) 1045–1055. https://doi.org/10.1016/J.CEJ.2019.03.032.
[68] Z. Hu, Z. Wen, Enhancing enzymatic digestibility of switchgrass by microwave-assisted alkali pretreatment, Biochem Eng J 38 (2008) 369–378. https://doi.org/10.1016/j.bej.2007.08.001.
[69] D.R. Keshwani, J.J. Cheng, Microwave-based alkali pretreatment of switchgrass and coastal microwave-based alkali pretreatment of switchgrass and coastal bermudagrass for bioethanol production bermudagrass for bioethanol production, (2011). https://doi.org/10.1002/btpr.371.
[70] Z.T. Zhao, J. Ding, J.W. Pang, M.Y. Bao, G. Luo, B.Y. Wang, B.F. Liu, L.Y. Zhang, N.Q. Ren, S.S. Yang, Pretreatments of lignocellulosic biomass for biohydrogen biorefinery: Recent progress, techno-economic feasibility and prospectives, Crit Rev Environ Sci Technol 55 (2025) 1070–1096. https://doi.org/10.1080/10643389.2025.2484892.
[71] M.R. Atelge, A.E. Atabani, J.R. Banu, D. Krisa, M. Kaya, C. Eskicioglu, G. Kumar, C. Lee, Y. Yildiz, S. Unalan, R. Mohanasundaram, F. Duman, A critical review of pretreatment technologies to enhance anaerobic digestion and energy recovery, Fuel 270 (2020) 117494. https://doi.org/10.1016/J.FUEL.2020.117494.
[72] E.D. Albuquerque, F.A.G. Torres, A.A.R. Fernandes, P.M.B. Fernandes, Combined effects of high hydrostatic pressure and specific fungal cellulase improve coconut husk hydrolysis, Process Biochem 51 (2016) 1767–1775. https://doi.org/10.1016/J.PROCBIO.2016.07.010.
[73] Y. Zhang, P. Zhang, J. Guo, W. Ma, L. Xiao, Spectroscopic analysis and biodegradation potential study of dissolved organic matters in sewage sludge treated with high-pressure homogenization, Bioresour Technol 135 (2013) 616–621. https://doi.org/10.1016/J.BIORTECH.2012.09.034.
[74] S.C.T. Oliveira, A.B. Figueiredo, D.V. Evtuguin, J.A. Saraiva, High pressure treatment as a tool for engineering of enzymatic reactions in cellulosic fibres, Bioresour Technol 107 (2012) 530–534. https://doi.org/10.1016/j.biortech.2011.12.093.
[75] S. Jin, G. Zhang, P. Zhang, S. Fan, F. Li, High-pressure homogenization pretreatment of four different lignocellulosic biomass for enhancing enzymatic digestibility, Bioresour Technol 181 (2015) 270–274. https://doi.org/10.1016/j.biortech.2015.01.069.
[76] E. Vorobiev, N. Lebovka, Applications of pulsed electric energy for biomass pretreatment in biorefinery, Biomass Fractionation Technologies for a Lignocellulosic Feedstock Based Biorefinery (2016) 151–168. https://doi.org/10.1016/B978-0-12-802323-5.00007-4.
[77] F.J. Barba, O. Parniakov, S.A. Pereira, A. Wiktor, N. Grimi, N. Boussetta, J.A. Saraiva, J. Raso, O. Martin-Belloso, D. Witrowa-Rajchert, N. Lebovka, E. Vorobiev, Current applications and new opportunities for the use of pulsed electric fields in food science and industry, Food Res Int 77 (2015) 773–798. https://doi.org/10.1016/J.FOODRES.2015.09.015.
[78] A.G. Pakhomov, D. Miklavcic, M.S. Markov, Advanced electroporation techniques in biology and medicine, Advanced Electroporation Techniques in Biology and Medicine (2010). https://doi.org/10.1201/EBK1439819067/ADVANCED-ELECTROPORATION-TECHNIQUES-BIOLOGY-MEDICINE-ANDREI-PAKHOMOV-DAMIJAN-MIKLAVCIC-MARKO-MARKOV.
[79] E. Vorobiev, N. Lebovka, Extraction from foods and biomaterials enhanced by pulsed electric energy, Innovative Food Processing Technologies: Extraction, Separation,

Component Modification and Process Intensification (2016) 31–56. https://doi.org/10.1016/B978-0-08-100294-0.00002-X.

[80] E. Vorobiev, N. Lebovka, Enhancing Extraction Processes in the Food Industry (2011) 25–83.

[81] P. Kumar, D.M. Barrett, M.J. Delwiche, P. Stroeve, Pulsed electric field pretreatment of switchgrass and wood chip species for biofuel production, Ind Eng Chem Res 50 (2011) 10996–11001. https://doi.org/10.1021/IE200555U.

[82] D. Carullo, G. Pataro, F. Donsì, G. Ferrari, Pulsed electric fields-assisted extraction of valuable compounds from arthrospira platensis: Effect of pulse polarity and mild heating, Front Bioeng Biotechnol 8 (2020) 551272. https://doi.org/10.3389/FBIOE.2020.551272/BIBTEX.

[83] M.B. Salerno, H.-S. Lee, P. Parameswaran, B.E. Rittmann, Using a pulsed electric field as a pretreatment for improved biosolids digestion and methanogenesis, Water Environ Res 81 (2009) 831–839. https://doi.org/10.2175/106143009x407366.

[84] S. Mahnič-Kalamiza, E. Vorobiev, D. Miklavčič, Electroporation in food processing and biorefinery, J Membr Biol247 (2014) 1279–1304. https://doi.org/10.1007/S00232-014-9737-X.

[85] Z.T. Zhao, J. Ding, J.W. Pang, M.Y. Bao, G. Luo, B.Y. Wang, B.F. Liu, L.Y. Zhang, N.Q. Ren, S.S. Yang, Pretreatments of lignocellulosic biomass for biohydrogen biorefinery: Recent progress, techno-economic feasibility and prospectives, Crit Rev Environ Sci Technol 55 (2025) 1070–1096. https://doi.org/10.1080/10643389.2025.2484892.

[86] V. Rooni, M. Raud, T. Kikas, The freezing pre-treatment of lignocellulosic material: A cheap alternative for Nordic countries, Energy 139 (2017) 1–7. https://doi.org/10.1016/J.ENERGY.2017.07.146.

[87] K.L. Chang, J. Thitikorn-amorn, J.F. Hsieh, B.M. Ou, S.H. Chen, K. Ratanakhanokchai, P.J. Huang, S.T. Chen, Enhanced enzymatic conversion with freeze pretreatment of rice straw, Biomass Bioenergy 35 (2011) 90–95. https://doi.org/10.1016/J.BIOMBIOE.2010.08.027.

[88] A.D. Henderson, B. Niblick, H.E. Golden, J.C. Bare, Modeling spatially resolved characterization factors for eutrophication potential in life cycle assessment, Int J Life Cycle Assess 26 (2021) 1832–1846. https://doi.org/10.1007/S11367-021-01956-4.

[89] S. Abinandan, K. Praveen, S.R. Subashchandrabose, K. Venkateswarlu, M. Megharaj, Life cycle assessment for the environmental sustainability of the immobilized acid-adapted microalgal technology in iron removal from acid mine drainage, ACS Sustain Chem Eng 8 (2020) 15670–15677. https://doi.org/10.1021/ACSSUSCHEMENG.0C05341/SUPPL_FILE/SC0C05341_SI_001.PDF.

[90] K.O. Olatunji, D.M. Madyira, J.O. Amos, Sustainable enhancement of biogas and methane yield of macroalgae biomass using different pretreatment techniques: A mini-review, Energy Environ 35 (2024) 1050–1088. https://doi.org/10.1177/0958305X231193869.

6 Thermal pretreatments

6.1 INTRODUCTION

The economic and social development of the world requires a continuous increase in global energy generation capacity. Currently, fossil fuels are the dominant source of energy, accounting for approximately 80% of the cumulative energy consumed. It is forecasted that energy demand will increase by about 35% between 2014 and 2035 [1]. Fossil fuels, as the primary source of energy, have caused different environmental challenges, and continuing to use this source of energy will lead to further damage to the ecosystem. Carbon neutrality is a crucial means of mitigating the environmental challenges caused by the extensive use of fossil fuels. Sourcing alternative energy sources with low carbon emissions is important for achieving net-zero emissions and energy security [2]. Different forms of renewable energy with minimal carbon footprints have been identified and utilized globally, including biomass, wind, solar, geothermal, and tidal energy [3,4]. The concept of biomass utilization in biorefineries for energy generation has been identified for its social, economic, and environmental impact on the economy, particularly in the context of the circular economy approach [5]. Biorefinery development enables the conversion of inedible renewable biomass into chemicals that replace fossil fuels and petroleum-based products, thereby promoting a sustainable environment. There has been a recent increase in the global acceptance of biofuel in recent years due to its environmental benefits [6]. Lignocellulose biomass is a readily available and economical feedstock to produce second-generation biofuels. It is a relatively inexpensive, renewable, and non-food feedstock that holds more promise for biorefinery. This feedstock can be sourced from agricultural and forest residues, organic fractions of municipal solid waste, energy crops, livestock manure, wastewater sludge, and algae [7,8]. The main components of lignocellulose biomass are lignin, cellulose, and hemicellulose, and the percentage of each constituent depends on the type, age, season, and geographical location of the biomass [9]. This composition of lignocellulose biomass makes it a unique material in energy and chemical production. Hemicellulose and cellulose are bound together by hydrogen and covalent bonds, and they are intertwined with the lignin polymer, which contributes to its recalcitrant properties [10]. The major challenge in applying biomass in biorefineries is its recalcitrant properties, which hinder conversion and result in low efficiency [11]. Therefore, pretreatment application is required to enhance the accessibility of organic matter in the biomass by breaking down the morphological arrangement of the fragmented biomass into its constituent components, including lignin, hemicellulose, and cellulose. Various pretreatment techniques, including thermal, biological, mechanical, chemical, and combined methods, have been developed to enhance the efficiency of bioenergy production. The primary aim of these techniques is to disrupt the microscopic, submicroscopic, and macroscopic arrangement of the biomass, making it accessible during downstream bioprocessing [12]. Thermal pretreatment is the application of

DOI: 10.1201/9781003633815-9

heat to solubilize the recalcitrant properties of biomass. The introduction of higher temperatures to biomass reduces its viscosity and solubilizes the biomass, making the organic matter more accessible [13]. The heat applied degrades the hydrogen bonds between the biomass molecules. The temperature in thermal pretreatment varies, ranging from 50 to 200 °C and above, depending on the type of biomass. One of the major challenges of thermal pretreatment is the tendency to generate inhibitory compounds at higher temperatures. There are different types of thermal pretreatment, each with its merits and limitations. This chapter discusses the various types of thermal pretreatment techniques and their influence on lignocellulose biomass and downstream bioprocessing. The environmental and social impacts of thermal pretreatment were discussed, and the recommendations for further studies were presented.

6.2 HYDROTHERMAL

This pretreatment technique subjected moist lignocellulose biomass to a higher temperature for a specified period to degrade its recalcitrant properties. During this technique, the applied temperature can reach up to 400 °C and is conducted at high pressure. Hydrothermal pretreatment can be categorized into three main processes: hydrothermal gasification, hydrothermal liquefaction, and hydrothermal carbonization [14], as illustrated in Figure 6.1. It is one of the most advanced pretreatment techniques used in biomass and has been applied in several biorefinery upscaling

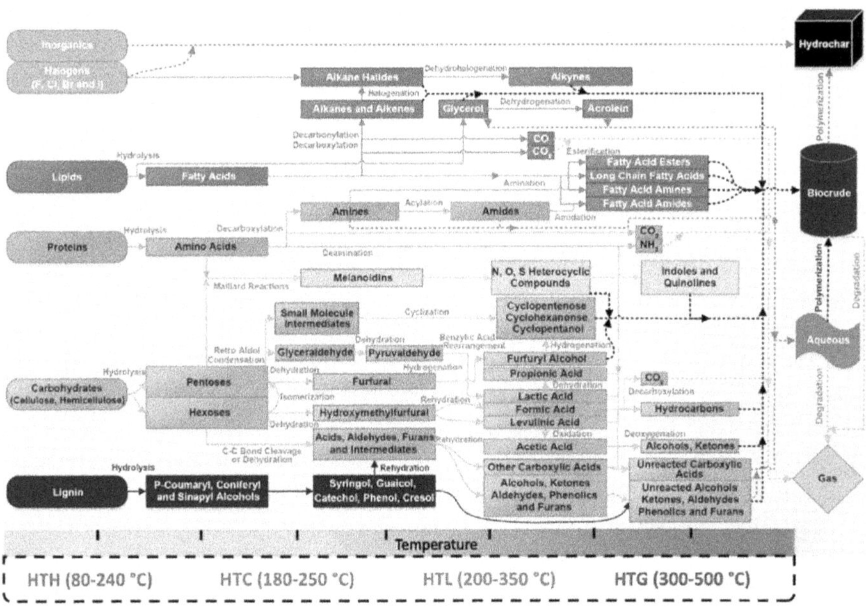

FIGURE 6.1 Hydrothermal pretreatment of lignocellulose biomass for bioenergy production [21].

approaches [15]. The technique uses water or steam, which leads to water auto-ionization into OH− and H_3O^+ at higher temperatures. The water then acts as a catalyst, facilitating the generation of acetate from xylan and the hydrolysis of glycosidic bonds in the xylan. Furthermore, the reduction of the pH of the medium serves as a catalyst for the hydrolysis reactions [16,17]. These actions stimulate xylan (hemicellulose) depolymerization but have a mild impact on the hydrolysis of lignin and cellulose [18]. The performance of hydrothermal pretreatment on hemicellulose solubilization and depolymerization depends on the process parameters, especially exposure time and temperature [12]. Pretreatment at higher temperatures and longer exposure times has been observed to release high concentrations of inhibitory compounds, such as 5-hydroxymethylfurfural and furfural, that can negatively affect the downstream process of biorefinery [19]. The ability of the process to pretreat biomass with up to 70% wt without the need to first dry it is a major advantage of the technique. The technique is applied in an aqueous system, and the available moisture in the biomass serves as a catalyst, solvent, reactant, or catalyst precursor [20].

Auto-catalyzed hydrothermal pretreatment is the easiest technique, whereby biomass is broken down in water heated to a temperature between 150 and 230 °C. The major benefits of this method are that no chemical additives (except water and alkali for further pH controls) are required, and no special anticorrosive equipment is required, which lowers the investment cost. Additionally, it can be carried out continuously at a high solid loading ratio, making it suitable for commercial applications [22]. Hydrothermal carbonization is another promising and sustainable hydrothermal technique. This technique is carried out at an autogenous pressure of 180–240 °C. It is most suitable when pretreating biomass with high moisture content, as it assists in reducing the biomass's oxygen, hydrogen, and moisture content through decarboxylation and dehydration [23]. The performance of this method depends on several key input parameters, including exposure time, temperature, pressure, and water supply. The primary product from this method is the hydrochar. Orange waste was pretreated using a hydrothermal technique with a 1:10 (w/w) ratio at different temperatures (100, 140, and 180 °C) and exposure times (30, 60, and 120 min). It was observed that hydrothermal pretreatment at 140 °C for 30 min produced optimum hydrolysis and fermentation. The fermentation and anaerobic digestion of residues released 42.3 g of biobutanol, 13.4 g of ethanol, 33.1 g of acetone, 104.5 l of biohydrogen, and 28.3 l of biomethane per kg of orange waste that has 4560 kJ of energy [24]. Hydrothermal pretreatment of wheat straw was carried out at 120, 140, 160, and 180 °C for 60 min. It was observed that pretreatment has a significant influence on the feedstock at temperatures above 160 °C, and the optimum biogas yield was improved by 53%, with a 69% reduction in volatile solids, at 180 °C, compared to the untreated feedstock [25]. It was observed that lignocellulose morphological disruption and increased cellulose content during pretreatment contributed to the improvement in biogas yield. Hydrothermal pretreatment of fresh vinegar residue was performed using 4.87 kW and 0.8 MPa for 30 min at 80, 120, and 160 °C, respectively. Optimum methane yield of 258.38 l/kg VS was recorded at 160 °C for 30 min [26]. Methane yield from Safflower straw was improved by 98.3% when pretreated with the hydrothermal technique at 120 °C for 60 min [27]. Despite the benefits of hydrothermal pretreatment, including the potential to release inhibitory compounds at higher temperatures, partial lignin degradation, and the need for hydrolysate removal that reduces sugar yield, these limitations necessitate further study.

6.3 WET OXIDATION

Wet oxidation pretreatment is a thermal pretreatment technique that utilizes gaseous oxygen as an oxidizing agent to pretreat biomass at elevated temperatures and pressures. The hydrolase process occurs at low temperature, while the oxidative reaction occurs at higher temperatures during wet oxidation pretreatments [10]. It is an exothermic reaction that minimizes the energy required and can effectively remove lignin, releasing lower inhibitory compounds. It was reported to effectively degrade biomass into cellulose-rich content and a liquid content consisting of solubilized hemicellulose sugars that can be used as raw materials during the downstream bioenergy process [28]. The process typically occurs at high temperatures ranging from 150 to 325 °C and high oxygen pressures of 10 to 200 bars. At high temperatures, biomass dissolution occurs, and high pressure supports the dissolution of oxidants. This process oxidizes lignin and hemicellulose, which are degraded into smaller pentose monomers with minimal influence on cellulose. Oxidant dose, oxygen pressure, temperature, and reaction time are the primary input parameters that determine the efficiency of this technique [29]. The process utilizes molecular oxygen in a liquid state to generate radicals that are used in the reaction and mineralization of the biomass, as shown in Equations 6.1 and 6.2 [30].

$$RH + O_2 \rightarrow R^* + HO_2^-, \tag{6.1}$$

$$RH + HO_2^- \rightarrow R^* + H_2O_2. \tag{6.2}$$

Radicals generated can react with additional oxygen and biomass to extend the chain or release various types of catalysts, as shown in Equations 6.3 and 6.4.

$$R^* + O_2 \rightarrow ROO^*, \tag{6.3}$$

$$R_1OO^* + R_2H \rightarrow R_1OOH + R_2^*. \tag{6.4}$$

The technique is also useful to reduce excess sludge or pretreatment of other bioenergy feedstocks, such as domestic, municipal, and industrial waste sludge [31]. Other important products, such as acetic acid, can be released during the wet oxidation pretreatment [32]. The addition of chemical solvents like sodium carbonate and alkaline peroxide to the process can enhance the disruption of hemicellulose, reduce process temperature, and minimize the release of inhibitory compounds, such as furfuraldehyde and furfurals [33].

Effluent from molasses-based distilleries undergoing biomethanation was pretreated with wet oxidation at temperatures of 150–200 °C for 15–120 min before entering the biorefinery. Pretreatment was observed to reduce the COD by around 54.75% and favors biogas production with up to 64% methane content [34]. A mixture of wheat straw and rice husk in a 1:1 ratio was pretreated with wet oxide using an air pressure of 5–10 bar and different temperatures. Pretreatment using 5 bars for 10 min at 195 °C was observed to improve enzymatic efficiency to 63 compared to 19% for the untreated feedstock. The treatment conditions also result in 52% lignin removal and 67% hemicellulose solubilization [35]. Three (3) biomasses of cassava peels, water hyacinth, and Ulva algae were subjected to wet oxidation pretreatment under two conditions. Water hyacinth was observed to produce optimal carbohydrates of 1.5 g/l, which corresponds to 95% efficiency, when pretreated at 1.5 bar, 125 °C for 45 min, and a pH of 3. However, optimum carbohydrate levels of 2.3 and 1.83 g/l were reported for cassava peels and Ulva algae at 1.8 bar, 130 °C for 75 min, and a pH of 1. The pretreatment efficiency for cassava peels and Ulva algae was 63 and 95%, respectively. At the optimum conditions, the ethanol released was 0.2, 0.26, and 0.31 g/l for water hyacinth, Ulva algae, and cassava peels, respectively [36]. The technique is an effective means of improving the available surface area of biomass for downstream bioprocessing through cellulose solubilization. Nonetheless, technology still faces some challenges that hinder its acceptance. The process requires a large quantity of oxygen, which increases the cost and is a major limitation in industrial-scale applications. Further studies are needed to investigate methods for reducing the oxygen consumed during the process, thereby reducing production costs. Investigating the production of other value-added products, such as volatile acids, is necessary to optimize the process.

6.4 LIQUID HOT WATER

Liquid hot water pretreatment, also known as hot compressed water, utilizes high temperatures and pressures (approximately 5 MPa) to maintain the water in a liquid state. Different researchers have referred to the technique by various names, including aqueous fractionation, hydrothermolysis, solvolysis, and aquasol [37]. The process is initiated by the hydro ions (H_3O^+) produced through the dissociation of water molecules. The technique does not need quick decomposition and addition of a chemical or a catalyst [38]. The process typically occurs at a temperature between 170 and 230 °C and a pressure of below 5 MPa [39]. The biomass remains in contact with water throughout the pretreatment period. Liquid hot water can be carried out in three ways, depending on the type of biomass and the direction of water flow into the reactor. In the flow-through method, hot water is allowed to flow through the biomass, which behaves as a stationary bed, and hydrolyzes and degrades the biomass components, removing them from the system. The second method is co-current; in this case, the biomass slurry and water are heated to a high temperature, held for a certain exposure time, and then released into a cool environment. The last method is countercurrent, where hot water is pumped against the biomass at controlled conditions [40]. Regardless of the method used, the process disrupts the biomass's morphological arrangement, leading to hemicellulose hydrolysis and the elimination of a small percentage of lignin, and makes cellulose more accessible for subsequent downstream processes, thereby minimizing the generation of process inhibitors that are released at higher temperatures [41]. The process required pH maintenance between 4 and 7, as the hemicellulose that can dissolve at this pH primarily consists of oligomeric content. Consequently, the generation of monosaccharides and further degradation materials, which can subsequently catalyze cellulose hydrolysis, is reduced [42]. The findings show that temperature and treatment time are significant for hemicellulosic sugar recovery and bioenergy release in the subsequent biorefinery [42–44]. To improve the efficiency of the technique, a chemical or a catalyst can be added or combined with another pretreatment technique.

Liquid hot water pretreatment was applied to poplar chips to reduce their recalcitrance characteristics before bioenergy production. The sample was treated in a fluidized sand bath at 180 °C for exposure times of 18, 28, 44, 56, and 70 min, with a ramping time of 39 min from room temperature to 180 °C. Results show a significant xylan solubilization of 50–77% (w/w dry matter), with remaining hemicellulose reduced by 63–75%. It was observed that pretreatment had a moderate influence on cellulose arrangement, resulting in a 6–9% improvement in crystallinity index, while the degree of polymerization was reduced by 35–65% [45]. Corn stalks were pretreated under a set of conditions using temperatures ranging from 120 to 180 °C and varying exposure times of 15 to 45 min. Different catalysts using sulfuric acid, formic acid, and phosphoric acid, with different concentrations of 0, 0.5, and 1.0% v/v, were added. An optimum glucose yield of 91.05% was recorded at 162.4 °C with 0.45% v/v sulfuric acid for 29.5 min. Morphological analysis of the sample using a scanning electron microscope shows that pretreatment alters the arrangement, and functional group analysis with Fourier transform infrared spectroscopy indicates a disruption in the functional groups [46]. Generally, this technique is promising from a cost-effective perspective, as it requires no chemicals and no special or corrosion-resistant equipment for the hydrolysis reaction. Due to the high volume of water needed, there is a lower concentration of solubilized lignin and hemicellulose release, which is another major benefit of this method. Compared to steam explosion pretreatment, this method released minimal inhibitory compounds with higher pentose recovery. This technique is easy, and the application of low temperatures that minimize biomass degradation eliminates the need for post-pretreatment neutralization or washing. Despite the many benefits of this technique, there are some limitations that limit its effectiveness. For instance, the method is not yet widely adopted at an industrial scale due to higher water requirements and high energy demand. Another challenge is the downstream processing because of the higher quantity of solubilized biomass with a lower concentration of solubilized products compared to other methods. High lignin solubility affects the recovery of hemicellulose from the solution. For this method to be economical and acceptable, further studies are required to overcome the identified challenges [47].

6.5 STEAM EXPLOSION

Steam explosion pretreatment is a thermos-pressure technique, similar to autohydrolysis, primarily because it is environmentally friendly. This technique utilizes saturated steam to heat the biomass at high pressure for a short duration, followed by the sudden release of pressure. This resulted in the expansion of the steam in the biomass, causing the disruption of individual bonds through degradation of the cell wall arrangement and solubilization of lignin and hemicellulose [48]. It significantly affects the physicochemical characteristics of biomass by degrading the microstructural arrangement, redistributing lignin, and hydrolyzing the hemicellulose content. The process consists of two phases: a chemical phase and a physical phase. The first phase is primarily chemical; at this phase, the biomass is subjected to high-temperature and pressure-saturated steam under hydrothermal conditions. The second phase is physical, where an explosive decompression denaturizes the matrix, rearranges the lignin, and reduces the particle size [48–50]. The biomass modification using steam pretreatment enhances the binding due to several factors. During pretreatment, hemicellulose is degraded and depolymerized through autohydrolysis, forming soluble sugars and other products that can be further repolymerized to become highly branched polymers during densification [51]. Additionally, high temperatures lead to depolymerization and thermal softening of lignin in the cell wall, causing realignment of the biomass matrix [52]. Lignin content will also escape from the lignocellulose matrix to the surface of the fibers, where reaction with moisture will behave as a plasticizer and bind during densification. The burster degradation reduces the particle size and rearranges the lignin on the particle surface, resulting in improved surface area and improved lignin availability, respectively [50]. The performance of steam explosion pretreatment depends on the temperature, exposure time, feedstock type, and particle size, and can be enhanced by adding chemicals like alkali or acid [53].

Steam explosion pretreatment has been experimented with different lignocellulosic biomass, and an improvement in bioenergy yield has been reported. For instance, agave bagasse was pretreated using absolute pressures of 0.38, 0.68, and 0.98 MPa for 2 min 48 s, 6 min, and 22 min at temperatures of 142, 164, and 179 °C. It was observed that an extract with a COD concentration of 32 g/l was achieved when the substrate was pretreated with 0.98 MPa for 22 min at 179 °C. Biomethane potential testing also shows an 89% COD removal, with a specific biomethane release of 290 ml/g COD [54]. Steam explosion pretreatment was applied to pulverized corncob pre-impregnated for 12 h in 0.5% wt H_2SO_4 using a solid-to-liquid ratio of 1:15. The sample was treated at 180 °C for 10 min under 1.0 MPa, and both the pretreated and untreated samples were analyzed. It was noticed that the carboxyl formation, lignin cleavage, and β-O-4′ cleavage were significantly disrupted after pretreatment. Although the sample still has an acceptable quantity of β-O-4′ bonds, which makes it useful for the subsequent downstream biorefinery. Lignin produced during the process has about 99% purity with a yield of 57.3% [55]. Biofibers from digested manure pretreated with steam explosion were reported to increase the biomethane yield by 67% compared to the untreated sample [56]. Results from the steam explosion pretreatment have shown that low temperatures and longer exposure times (e.g., 190 °C for 10 min) produced a better result than higher temperatures for shorter exposure times (e.g., 270 °C for 1 min) because lower inhibitory compounds are released at lower temperatures. The merits of this technique include its ability to degrade biomass without chemical addition, its environmental friendliness compared to other techniques, quick decompression, which produces fast and efficient sugar generation, and the production of lower inhibitory compounds, thereby reducing the requirement for subsequent detoxification [57]. However, this technique is facing some drawbacks that limit its application. The energy required to maintain high pressure and temperature is high, which can make the process uneconomical.

The possibility of sugar degradation can result in the generation of biofuel-inhibitory materials, such as furan derivatives and aromatic compounds [58]. The method partially degrades the lignin, which is a major limitation when applied to biomass with high lignin content. Despite the wider application of this pretreatment method, process parameter control is not yet fully mastered, and the detailed mechanisms underlying it are still poorly understood.

6.6 CONVENTIONAL/AUTOCLAVE

Autoclave pretreatment is a conventional thermal pretreatment that combines high temperature and pressure to effect changes and catalyze the hydrolysis of covalent bonds in the biomass's morphological arrangement [59]. This technique reduces the energy required and wastewater released during thermal pretreatment, making it one of the most promising techniques. It is one of the most suitable thermal pretreatment techniques, which utilizes moist heat under pressure from an autoclave. In comparison with other thermal pretreatment techniques, this method supplies both heat and stress simultaneously, making the biomass's recalcitrant properties vulnerable to chemical reactions. The air moisture helps spread the heat uniformly within the autoclave chamber, thereby enhancing the efficiency of the heating process [60]. The samples are exposed to moist hot air, and the heat is transferred by conduction and convection. In this technology, moist biomass inside the pressure vessel is heated to 120–180 °C, resulting in a vapothermal environment in the reactor and achieving a pressurized temperature treatment. At the end of the exposure time, the pressure built up in the pressure vessel is quickly released, which will produce a form of steam-explosion influence on the biomass [61]. During the rapid decompression resulting from the quick withdrawal of pressure, the liquid in the biomass matrix suddenly expands, leading to a large shearing force that breaks down the cell arrangement. When compared with steam explosion, the pressure is mostly distributed evenly across the biomass due to a lower pressure level and deflation. Therefore, the forces applied to the biomass matrix are much weaker [62]. The efficiency of autoclave pretreatment depends on the type of biomass and various process parameters, including rotation speed, treatment temperature, pressure, and exposure time. Autoclave pretreatment was applied to *Ageratum conyzoides* biomass at 90 °C for 90 min. The results showed a simultaneous increase in solubilization and soluble chemical oxygen demand, which were 54.59 and 55.33%, respectively. The total biomethane released was 4053 ml CH_4/g VS in 35 days, compared to 3011 ml CH_4/g VS from untreated samples in 50 days [63]. Autoclave pretreatment at varying temperatures of 90, 100, 110, and 120 °C was conducted on Arachis hypogea shells for durations of 30 and 60 min before anaerobic digestion. Scanning electron microscope analysis of the samples revealed that pretreatment increased porosity, which was further enhanced with increasing temperature and time. It was observed from Fourier transform infrared spectroscopy that autoclave pretreatment removed hemicellulose and lignin. An optimum biomethane yield of 23.96% increase compared to untreated feedstock was recorded when a temperature of 100 °C was used for 30 min. This study reported that temperatures and time beyond the optimum conditions reduce the biomethane yield [64]. The merits of this method include its high-pressure reaction, which prevents dehydration of the biomass at higher temperatures and generates low-toxic compounds, thereby further increasing biorefinery efficiency [59].

6.7 HOT AIR OVEN

Hot air oven biomass pretreatment is a technique that utilizes hot air to disrupt the recalcitrant properties of lignocellulose biomass. The mode of heat transfer in this method is by convection, followed by conduction. A fan is used to circulate the dry heat uniformly within the chamber. The dry circulated in the chamber is sucked by the outer layer of the biomass and then moves gradually until it reaches the center of the biomass, and the cell wall of the biomass is ruptured during this process. The movement of hot air through the container containing the feedstock is a form of convection, and the transfer from the container to the feedstock is a form of conduction [65]. The applied hot air required will be at atmospheric pressure and can be sourced from waste heat or recovered from other sources, such as process integration with combined heat and power (CHP) settings or recovery from a hot water boiler [66]. Hot air pretreatment at low temperatures of 75–85 °C was observed to have similar effects to biomass pretreated with autoclave pretreatment at 121 °C. Hot air generation and application are cheaper and easier compared to steam autoclaving, which needs 60% additional energy than 85 °C hot air pretreatment, with the additional cost of pressurized equipment [67]. The major difference between the hot air and autoclave pretreatment is that hot air can be applied at normal pressure in an open environment, whereas autoclave pretreatment requires a pressurized steam environment [68]. This method significantly improves energy efficiency, is economical, and is environmentally benign for the integrated production of food and biofuel. However, the influence of hot air pretreatment on biomass pretreatment is poorly understood. The comparison of the efficiency of hot air and saturated steam pretreatment on lignin and carbohydrate disruption is still unknown. The information about the dynamics of biomass composition after hot air pretreatment is still scanty.

6.8 ENVIRONMENTAL AND SOCIAL IMPACTS

The use of thermal pretreatment of biomass in bioenergy production has several environmental benefits, including increasing the availability of biomass for downstream processes and reducing dependence on fossil fuels and their associated greenhouse gas emissions. Technology also enhances the recovery and valorization of biomass products, enabling the reduction of waste and the production of useful by-products. Some thermal pretreatment techniques, such as steam explosion, hot air, and liquid hot water, are regarded as eco-benign because they are economical and utilize water instead of chemicals [69]. Hydrothermal pretreatment utilizes water as a reactant, solvent, or catalyst, and does not require washing or drying the wet biomass before downstream operations. This approach is environmentally friendly, as it releases no chemicals or wastewater into the environment. Thermal pretreatment takes place at an elevated temperature, which tends to produce inhibitory compounds. The management of inhibitory materials, such as 5-HMF, phenolic compounds, furfural, and acetic acids produced during pretreatment, is another major challenge that can impact the environment when the biomass conversion rate is reduced, thereby increasing the amount of waste generated during the downstream process [70]. Thermal pretreatment is generally considered eco-friendly; however, the process can lead to the release of carbon dioxide due to the high energy required. Economic feasibility is important for the commercial production of bioenergy, and pretreatment of biomass is one of the major stages that is expensive in biorefinery. The energy required during thermal pretreatment is high, and considering the cost of energy, it increases the production cost, which hinders its competitiveness with fossil fuels. The process efficiency and environmental concerns can be reduced by utilizing various energy recovery methods, such as CHP in biorefineries. Integrating energy sources with a lower carbon footprint during thermal pretreatment is crucial for mitigating the environmental impact of this process [71].

6.9 CONCLUSION

An efficient and environmentally friendly biomass pretreatment technique has been identified as a concept for sustainable biorefineries. An overview of thermal pretreatment reveals that it is a promising technique that can disrupt the recalcitrant properties of biomass, enhance cellulose accessibility, and produce optimal bioenergy during the downstream biorefinery process while also releasing other value-added products. This study reviews recent developments in thermal pretreatment techniques, identifying their merits and limitations. It was noticed that many of the techniques have gradually improved in their strength to disrupt the microstructural arrangement of biomass. Nonetheless, a thorough review of the techniques reveals that further consistent efforts are still necessary to develop more efficient, economical, and environmentally benign thermal pretreatment techniques that can enhance the delignification of biomass for biorefinery applications. Findings from this study indicate that sustainability, energy savings, capital cost reduction, and downstream biorefinery efficiency remain limitations to the commercialization of thermal biomass pretreatment. Due to the high temperature required in thermal pretreatment, the release of inhibitory compounds such as furaldehydes, aliphatic acids, phenol, vanillic acid, formaldehyde, 4-hydroxybenzoic acid, uronic acid, and cinnamaldehyde has a negative impact on downstream bioprocessing. Therefore, future studies should focus more on gaining a sound understanding of the thermal pretreatment process mechanism to provide a tailor-made process that considers the various biomass compositions and the expected end products. Additionally, further investigations are needed to determine the most efficient method and its feasibility for identifying the possibility of applying it on an industrial scale in a biorefinery. This study will support the development of energy-efficient, environmentally friendly, and cost-effective pretreatment methods and enhance the current thermal pretreatment methods by providing solutions to their limitations.

REFERENCES

[1] T. Ahmad, D. Zhang, A critical review of comparative global historical energy consumption and future demand: The story told so far, Energy Rep 6 (2020) 1973–1991. https://doi.org/10.1016/J.EGYR.2020.07.020.

[2] Q. Hassan, P. Viktor, T.J. Al-Musawi, B. Mahmood Ali, S. Algburi, H.M. Alzoubi, A. Khudhair Al-Jiboory, A. Zuhair Sameen, H.M. Salman, M. Jaszczur, The renewable energy role in the global energy transformations, Renew Energy Focus 48 (2024) 100545. https://doi.org/10.1016/J.REF.2024.100545.

[3] F. Mutua, C. Bess, A. Mumbo, Renewable energy: The paradigm shift-energy, development and climate change, Kenya—South Africa context, J Eng Agric Environ 6 (2021). https://doi.org/10.37017/JEAE-VOLUME6-NO2.2020-2.

[4] M.M. Vanegas Cantarero, Of renewable energy, energy democracy, and sustainable development: A roadmap to accelerate the energy transition in developing countries, Energy Res Soc Sci 70 (2020) 101716. https://doi.org/10.1016/J.ERSS.2020.101716.

[5] P. Morone, G. Yilan, A paradigm shift in sustainability: From lines to circles, Acta Innov (2020) 5–16. https://doi.org/10.32933/ACTAINNOVATIONS.36.1.

[6] Y. Sewsynker-Sukai, E.B. Gueguim Kana, Simultaneous saccharification and bioethanol production from corn cobs: Process optimization and kinetic studies, Bioresour Technol 262 (2018) 32–41. https://doi.org/10.1016/J.BIORTECH.2018.04.056.

[7] G.I. Kulichkova, T.S. Ivanova, M. Köttner, O.I. Volodko, S.I. Spivak, S.P. Tsygankov, Y.B. Blume, Plant feedstocks and their biogas production potentials, Open Agric J 14 (2020) 219–234. https://doi.org/10.2174/1874331502014010219.

[8] Y. Wang, J. Wang, J. Schuler, D. Hartley, T. Volk, M. Eisenbies, Optimization of harvest and logistics for multiple lignocellulosic biomass feedstocks in the northeastern United States, Energy 197 (2020) 117260. https://doi.org/10.1016/J.ENERGY.2020.117260.

[9] K.O. Olatunji, D.M. Madyira, Enhancing microstructural and biomethane yield of Xyris capensis using combined oxidative and nanoparticle additive pretreatment methods, Biomass Convers Biorefin 1 (2023) 1–15. https://doi.org/10.1007/S13399-023-04401-8/FIGURES/5.

[10] R. Banu J, S. Sugitha, S. Kavitha, Y. Kannah R, J. Merrylin, G. Kumar, Lignocellulosic biomass pretreatment for enhanced bioenergy recovery: Effect of lignocelluloses recalcitrance and enhancement strategies, Front Energy Res 9 (2021) 679. https://doi.org/10.3389/FENRG.2021.646057/BIBTEX.

[11] C. Huang, X. Jiang, X. Shen, J. Hu, W. Tang, X. Wu, A. Ragauskas, H. Jameel, X. Meng, Q. Yong, Lignin-enzyme interaction: A roadblock for efficient enzymatic hydrolysis of lignocellulosics, Renew Sustain Energy Rev 154 (2022) 111822. https://doi.org/10.1016/J.RSER.2021.111822.

[12] K.O. Olatunji, N.A. Ahmed, O. Ogunkunle, Optimization of biogas yield from lignocellulosic materials with different pretreatment methods: A review, Biotechnol Biofuels 14 (2021) 1–34. https://doi.org/10.1186/S13068-021-02012-X.

[13] J.R. Banu, S. Kavitha, R.Y. Kannah, T.M.M. Usman, G. Kumar, Application of chemo thermal coupled sonic homogenization of marine macroalgal biomass for energy efficient volatile fatty acid recovery, Bioresour Technol 303 (2020) 122951. https://doi.org/10.1016/J.BIORTECH.2020.122951.

[14] R. Bao, S. Wang, J. Feng, Y. Duan, K. Liu, J. Zhao, H. Liu, J. Yang, A review of hydrothermal biomass liquefaction: Operating parameters, reaction mechanism, and bio-oil yields and compositions, Energy Fuels 38 (2024) 8437–8459. https://doi.org/10.1021/ACS.ENERGYFUELS.4C00240/ASSET/IMAGES/MEDIUM/EF4C00240_0012.GIF.

[15] H.A. Ruiz, M. Conrad, S.N. Sun, A. Sanchez, G.J.M. Rocha, A. Romaní, E. Castro, A. Torres, R.M. Rodríguez-Jasso, L.P. Andrade, I. Smirnova, R.C. Sun, A.S. Meyer, Engineering aspects of hydrothermal pretreatment: From batch to continuous operation, scale-up and pilot reactor under biorefinery concept, Bioresour Technol 299 (2020) 122685. https://doi.org/10.1016/J.BIORTECH.2019.122685.
[16] N.S. Kapu, Z. Yuan, X.F. Chang, R. Beatson, D.M. Martinez, H.L. Trajano, Insight into the evolution of the proton concentration during autohydrolysis and dilute-acid hydrolysis of hemicellulose, Biotechnol Biofuels 9 (2016) 1–10. https://doi.org/10.1186/S13068-016-0619-6/FIGURES/7.
[17] M. Pedersen, A. Viksø-Nielsen, A.S. Meyer, Monosaccharide yields and lignin removal from wheat straw in response to catalyst type and pH during mild thermal pretreatment, Process Biochemistry 45 (2010) 1181–1186. https://doi.org/10.1016/J.PROCBIO.2010.03.020.
[18] B.E. Morales-Contreras, N. Flórez-Fernández, M. Dolores Torres, H. Domínguez, R.M. Rodríguez-Jasso, H.A. Ruiz, Hydrothermal systems to obtain high value-added compounds from macroalgae for bioeconomy and biorefineries, Bioresour Technol 343 (2022) 126017. https://doi.org/10.1016/J.BIORTECH.2021.126017.
[19] X. Song, A.C. Wachemo, L. Zhang, T. Bai, X. Li, X. Zuo, H. Yuan, Effect of hydrothermal pretreatment severity on the pretreatment characteristics and anaerobic digestion performance of corn stover, Bioresour Technol 289 (2019) 121646. https://doi.org/10.1016/J.BIORTECH.2019.121646.
[20] A. Kruse, N. Dahmen, Water—A magic solvent for biomass conversion, J Supercrit Fluids 96 (2015) 36–45. https://doi.org/10.1016/J.SUPFLU.2014.09.038.
[21] Y. Zhou, J. Remón, X. Pang, Z. Jiang, H. Liu, W. Ding, Hydrothermal conversion of biomass to fuels, chemicals and materials: A review holistically connecting product properties and marketable applications, Sci Total Environ 886 (2023) 163920. https://doi.org/10.1016/J.SCITOTENV.2023.163920.
[22] T. Scapini, M.S.N. dos Santos, C. Bonatto, J.H.C. Wancura, J. Mulinari, A.F. Camargo, N. Klanovicz, G.L. Zabot, M.V. Tres, G. Fongaro, H. Treichel, Hydrothermal pretreatment of lignocellulosic biomass for hemicellulose recovery, Bioresour Technol 342 (2021) 126033. https://doi.org/10.1016/J.BIORTECH.2021.126033.
[23] I. Saif, N. Thakur, P. Zhang, L. Zhang, X. Xing, J. Yue, Z. Song, L. Nan, S. Yujun, M. Usman, E.S. Salama, X. Li, Biochar assisted anaerobic digestion for biomethane production: Microbial symbiosis and electron transfer, J Environ Chem Eng 10 (2022) 107960. https://doi.org/10.1016/J.JECE.2022.107960.
[24] F. Saadatinavaz, K. Karimi, J.F.M. Denayer, Hydrothermal pretreatment: An efficient process for improvement of biobutanol, biohydrogen, and biogas production from orange waste via a biorefinery approach, Bioresour Technol 341 (2021) 125834. https://doi.org/10.1016/J.BIORTECH.2021.125834.
[25] A.A. Rajput, Zeshan, C. Visvanathan, Effect of thermal pretreatment on chemical composition, physical structure and biogas production kinetics of wheat straw, J Environ Manage 221 (2018) 45–52. https://doi.org/10.1016/j.jenvman.2018.05.011.
[26] G. Ran, D. Li, T. Zheng, X. Liu, L. Chen, Q. Cao, Z. Yan, Hydrothermal pretreatment on the anaerobic digestion of washed vinegar residue, Bioresour Technol 248 (2018) 265–271. https://doi.org/10.1016/J.BIORTECH.2017.06.068.
[27] S.S. Hashemi, K. Karimi, S. Mirmohamadsadeghi, Hydrothermal pretreatment of safflower straw to enhance biogas production, Energy 172 (2019) 545–554. https://doi.org/10.1016/j.energy.2019.01.149.
[28] P. Alvira, E. Tomás-Pejó, M. Ballesteros, M.J. Negro, Pretreatment technologies for an efficient bioethanol production process based on enzymatic hydrolysis: A review, Bioresour Technol 101 (2010) 4851–4861. https://doi.org/10.1016/j.biortech.2009.11.093.

[29] J. Chung, M. Lee, J. Ahn, W. Bae, Y.W. Lee, H. Shim, Effects of operational conditions on sludge degradation and organic acids formation in low-critical wet air oxidation, J Hazard Mater 162 (2009) 10–16. https://doi.org/10.1016/J.JHAZMAT.2008.05.038.
[30] M.M. M'Arimi, C.A. Mecha, A.K. Kiprop, R. Ramkat, Recent trends in applications of advanced oxidation processes (AOPs) in bioenergy production: Review, Renew Sustain Energy Rev 121 (2020) 109669. https://doi.org/10.1016/J.RSER.2019.109669.
[31] G. Bertanza, R. Galessi, L. Menoni, R. Salvetti, E. Slavik, S. Zanaboni, Wet oxidation of sewage sludge: Full-scale experience and process modeling, Environ Sci Pollut Res 22 (2015) 7306–7316. https://doi.org/10.1007/S11356-014-3144-9/FIGURES/9.
[32] K. Hii, S. Baroutian, R. Parthasarathy, D.J. Gapes, N. Eshtiaghi, A review of wet air oxidation and thermal hydrolysis technologies in sludge treatment, Bioresour Technol 155 (2014) 289–299. https://doi.org/10.1016/J.BIORTECH.2013.12.066.
[33] S. Banerjee, R. Sen, S. Mudliar, R.A. Pandey, T. Chakrabarti, D. Satpute, Alkaline peroxide assisted wet air oxidation pretreatment approach to enhance enzymatic convertibility of rice husk, Biotechnol Prog 27 (2011) 691–697. https://doi.org/10.1002/btpr.589.
[34] T. Sarat Chandra, S.N. Malik, G. Suvidha, M.L. Padmere, P. Shanmugam, S.N. Mudliar, Wet air oxidation pretreatment of biomethanated distillery effluent: Mapping pretreatment efficiency in terms color, toxicity reduction and biogas generation, Bioresour Technol 158 (2014) 135–140. https://doi.org/10.1016/J.BIORTECH.2014.01.106.
[35] A. Sharma, A. Ghosh, R.A. Pandey, S.N. Mudliar, Wet air oxidation pretreatment of mixed lignocellulosic biomass to enhance enzymatic convertibility, Korean Chem Eng Res 53 (2015) 216–223. https://doi.org/10.9713/KCER.2015.53.2.216.
[36] Y.S. Ahou, M.N.A. Christami, S. Awad, C.R. Priadi, L. Baba-Moussa, S.S. Moersidik, Y. Andres, Wet oxidation pretreatment effect for enhancing bioethanol production from cassava peels, water hyacinth, and green algae (Ulva), AIP Conf Proc 2255 (2020). https://doi.org/10.1063/5.0013560.
[37] X. Zhuang, W. Wang, Q. Yu, W. Qi, Q. Wang, X. Tan, G. Zhou, Z. Yuan, Liquid hot water pretreatment of lignocellulosic biomass for bioethanol production accompanying with high valuable products, Bioresour Technol 199 (2016) 68–75. https://doi.org/10.1016/J.BIORTECH.2015.08.051.
[38] A.K. Kumar, S. Sharma, Recent updates on different methods of pretreatment of lignocellulosic feedstocks: A review, Bioresour Bioprocess 4 (2017) 1–19. https://doi.org/10.1186/S40643-017-0137-9.
[39] Ó.J. Sánchez, C.A. Cardona, Trends in biotechnological production of fuel ethanol from different feedstocks, Bioresour Technol 99 (2008) 5270–5295. https://doi.org/10.1016/J.BIORTECH.2007.11.013.
[40] M.N. Aftab, I. Iqbal, F. Riaz, A. Karadag, M. Tabatabaei, M.N. Aftab, I. Iqbal, F. Riaz, A. Karadag, M. Tabatabaei, Different pretreatment methods of lignocellulosic biomass for use in biofuel production, Biomass for Bioenergy—Recent Trends and Future Challenges (2019). https://doi.org/10.5772/INTECHOPEN.84995.
[41] M.E. Vallejos, F.E. Felissia, J. Kruyeniski, M.C. Area, Kinetic study of the extraction of hemicellulosic carbohydrates from sugarcane bagasse by hot water treatment, Ind Crops Prod 67 (2015) 1–6. https://doi.org/10.1016/J.INDCROP.2014.12.058.
[42] C. Peral, Biomass pretreatment strategies (technologies, environmental performance, economic considerations, industrial implementation), Biotransformation of Agricultural Waste and By-Products: The Food, Feed, Fibre, Fuel (4F) Economy (2016) 125–160. https://doi.org/10.1016/B978-0-12-803622-8.00005-7.
[43] J.A. Pérez, I. Ballesteros, M. Ballesteros, F. Sáez, M.J. Negro, P. Manzanares, Optimizing liquid hot water pretreatment conditions to enhance sugar recovery from wheat straw

for fuel-ethanol production, Fuel 87 (2008) 3640–3647. https://doi.org/10.1016/J.FUEL.2008.06.009.

[44] N. Mosier, C. Wyman, B. Dale, R. Elander, Y.Y. Lee, M. Holtzapple, M. Ladisch, Features of promising technologies for pretreatment of lignocellulosic biomass, Bioresour Technol 96 (2005) 673–686. https://doi.org/10.1016/j.biortech.2004.06.025.

[45] M. Li, S. Cao, X. Meng, M. Studer, C.E. Wyman, A.J. Ragauskas, Y. Pu, The effect of liquid hot water pretreatment on the chemical-structural alteration and the reduced recalcitrance in poplar, Biotechnol Biofuels 10 (2017) 1–13. https://doi.org/10.1186/S13068-017-0926-6/SCHEMES/1.

[46] N. Suriyachai, K. Weerasai, S. Upajak, P. Khongchamnan, W. Wanmolee, N. Laosiripojana, V. Champreda, K. Suwannahong, S. Imman, Efficiency of catalytic liquid hot water pretreatment for conversion of corn stover to bioethanol, ACS Omega 5 (2020) 29872–29881. https://doi.org/10.1021/ACSOMEGA.0C04054/ASSET/IMAGES/MEDIUM/AO0C04054_M004.GIF.

[47] S. Imman, N. Laosiripojana, V. Champreda, Effects of liquid hot water pretreatment on enzymatic hydrolysis and physicochemical changes of corncobs, Appl Biochem Biotechnol 184 (2018) 432–443. https://doi.org/10.1007/S12010-017-2541-1.

[48] A. Shukla, D. Kumar, M. Girdhar, A. Kumar, A. Goyal, T. Malik, A. Mohan, Strategies of pretreatment of feedstocks for optimized bioethanol production: Distinct and integrated approaches, Biotechnol Biofuels Bioprod16 (2023) 1–33. https://doi.org/10.1186/S13068-023-02295-2.

[49] T.R. Sarker, F. Pattnaik, S. Nanda, A.K. Dalai, V. Meda, S. Naik, Hydrothermal pretreatment technologies for lignocellulosic biomass: A review of steam explosion and subcritical water hydrolysis, Chemosphere 284 (2021) 131372. https://doi.org/10.1016/J.CHEMOSPHERE.2021.131372.

[50] Y. Yu, J. Wu, X. Ren, A. Lau, H. Rezaei, M. Takada, X. Bi, S. Sokhansanj, Steam explosion of lignocellulosic biomass for multiple advanced bioenergy processes: A review, Renew Sustain Energy Rev 154 (2022) 111871. https://doi.org/10.1016/J.RSER.2021.111871.

[51] Q. Sun, W.J. Chen, B. Pang, Z. Sun, S.S. Lam, C. Sonne, T.Q. Yuan, Ultrastructural change in lignocellulosic biomass during hydrothermal pretreatment, Bioresour Technol 341 (2021) 125807. https://doi.org/10.1016/J.BIORTECH.2021.125807.

[52] S. Maitra, V. Singh, Balancing sugar recovery and inhibitor generation during energycane processing: Coupling cryogenic grinding with hydrothermal pretreatment at low temperatures, Bioresour Technol 321 (2021) 124424. https://doi.org/10.1016/J.BIORTECH.2020.124424.

[53] Y. Huang, X. Wei, S. Zhou, M. Liu, Y. Tu, A. Li, P. Chen, Y. Wang, X. Zhang, H. Tai, L. Peng, T. Xia, Steam explosion distinctively enhances biomass enzymatic saccharification of cotton stalks by largely reducing cellulose polymerization degree in G. barbadense and G. hirsutum, Bioresour Technol 181 (2015) 224–230. https://doi.org/10.1016/J.BIORTECH.2015.01.020.

[54] B. Weber, A. Estrada-Maya, A.C. Sandoval-Moctezuma, I.G. Martínez-Cienfuegos, Anaerobic digestion of extracts from steam exploded Agave tequilana bagasse, J Environ Manage 245 (2019) 489–495. https://doi.org/10.1016/J.JENVMAN.2019.05.093.

[55] X. Zhang, J. Zhu, L. Sun, Q. Yuan, G. Cheng, D.S. Argyropoulos, Extraction and characterization of lignin from corncob residue after acid-catalyzed steam explosion pretreatment, Ind Crops Prod 133 (2019) 241–249. https://doi.org/10.1016/J.INDCROP.2019.03.027.

[56] E. Bruni, A.P. Jensen, I. Angelidaki, Steam treatment of digested biofibers for increasing biogas production, Bioresour Technol 101 (2010) 7668–7671. https://doi.org/10.1016/j.biortech.2010.04.064.

[57] K.O. Olatunji, D.M. Madyira, J.O. Amos, Sustainable enhancement of biogas and methane yield of macroalgae biomass using different pretreatment techniques: A mini-review, Energy Environ 35 (2024) 1050–1088. https://doi.org/10.1177/0958305X231193869.
[58] A. Verardi, A. Blasi, T. Marino, A. Molino, V. Calabrò, Effect of steam-pretreatment combined with hydrogen peroxide on lignocellulosic agricultural wastes for bioethanol production: Analysis of derived sugars and other by-products, J Energy Chem 27 (2018) 535–543. https://doi.org/10.1016/J.JECHEM.2017.11.007.
[59] T. Hute, I.A. Sanusi, E.B.G. Kana, E.L. Meyer, Y. Sewsynker-Sukai, Comparative assessment of autoclave- and microwave-facilitated seawater pretreatments for the enhancement of sugar recovery from banana pseudostem, Biomass Convers Biorefin 15 (2025) 12565–12579. https://doi.org/10.1007/S13399-024-06231-8/TABLES/5.
[60] A. Hamdy, S.A. Elhafez, H. Hamad, R. Ali, The interplay of autoclaving with oxalate as pretreatment technique in the view of bioethanol production based on corn stover, Polymers 13 (2021) 3762. https://doi.org/10.3390/POLYM13213762.
[61] M. Scherzinger, M. Kaltschmitt, Heat induced pre-treatment technologies for lignocellulosic biomass. A comparison of different processes and techniques, J Ecol Eng 20 (2019) 134–146. https://doi.org/10.12911/22998993/109876.
[62] Z. Yu, B. Zhang, F. Yu, G. Xu, A. Song, A real explosion: The requirement of steam explosion pretreatment, Bioresour Technol 121 (2012) 335–341. https://doi.org/10.1016/J.BIORTECH.2012.06.055.
[63] B. Saha, M. Khwairakpam, A.S. Kalamdhad, Thermal pre-treatment—A prerequisite for the reduction of hydrolysis stage during anaerobic digestion of Ageratum conyzoides, Mater Sci Energy Technol 4 (2021) 34–45. https://doi.org/10.1016/J.MSET.2020.12.002.
[64] K.O. Olatunji, D.M. Madyira, N.A. Ahmed, O. Ogunkunle, Biomethane production from Arachis hypogea shells: Effect of thermal pretreatment on substrate structure and yield, Biomass Convers Biorefin 14 (2024) 6925–6938. https://doi.org/10.1007/S13399-022-02731-7/METRICS.
[65] V.B. Barua, A.S. Kalamdhad, Effect of various types of thermal pretreatment techniques on the hydrolysis, compositional analysis and characterization of water hyacinth, Bioresour Technol 227 (2017) 147–154. https://doi.org/10.1016/J.BIORTECH.2016.12.036.
[66] L. Eilertsen, M. Hultberg, N. Lee, P. Saroj, M. Swaine, F. Chen, S. Xiong, The microbial community in a substrate of solid-state fermentation by lentinula edodes: A preliminary study, Fermentation 9 (2023) 736. https://doi.org/10.3390/FERMENTATION9080736.
[67] M. Wei, S. Xiong, F. Chen, P. Geladi, L. Eilertsen, O. Myronycheva, T.A. Lestander, M. Thyrel, Energy smart hot-air pasteurisation as effective as energy intense autoclaving for fungal preprocessing of lignocellulose feedstock for bioethanol fuel production, Renew Energy 155 (2020) 237–247. https://doi.org/10.1016/J.RENENE.2020.03.154.
[68] P.J. Fellows, Food Processing Technology Principles and Practice. Fourth Edition, Woodhead Publishing Series in Food Science, Technology and Nutrition (2017). http://www.sciencedirect.com:5070/book/9780081019078/food-processing-technology (accessed August 22, 2025).
[69] K.O. Olatunji, D.M. Madyira, N.A. Ahmed, O. Adeleke, O. Ogunkunle, Biomethane yield modeling and optimization from thermally pretreated Arachis hypogea shells using response surface methodology and artificial neural network, Bioresour Technol Rep 20 (2022). https://doi.org/10.1016/j.biteb.2022.101236.
[70] M. Mujtaba, L. Fernandes Fraceto, M. Fazeli, S. Mukherjee, S.M. Savassa, G. Araujo de Medeiros, A. do Espírito Santo Pereira, S.D. Mancini, J. Lipponen, F. Vilaplana,

Lignocellulosic biomass from agricultural waste to the circular economy: A review with focus on biofuels, biocomposites and bioplastics, J Clean Prod 402 (2023) 136815. https://doi.org/10.1016/J.JCLEPRO.2023.136815.

[71] N.K. Obiora, C.O. Ujah, C.O. Asadu, F.O. Kolawole, B.N. Ekwueme, Production of hydrogen energy from biomass: Prospects and challenges, Green Technol Sustain 2 (2024) 100100. https://doi.org/10.1016/J.GRETS.2024.100100.

7 Nanoparticle additive and combined pretreatment

7.1 INTRODUCTION

The degree of challenges posed by the excessive burning of fossil fuels on the climate has increased the need to source alternative sources of energy that can satisfy the demand for energy to support the new lifestyle resulting from civilization and technological advancement. Increasing the stream of clean, sustainable, safe, and environmentally friendly energy will be challenging and require innovative solutions and sufficient capital. Different technology and resource choices can minimize emissions from electricity production, shipping, heating/boiler, and transportation sectors, thereby strengthening energy security [1]. The biorefinery approach is a crucial path that can facilitate the transition from an oil-based economy to a promising bioeconomy, enabling more efficient and sustainable development through a bio-circular economy. The biorefinery approach aims to generate biofuels, bioproducts, and chemicals from readily available biomass [2,3]. Bioenergy production from lignocellulose biomass has been identified as a source of clean energy that can replace fossil fuels due to its availability, its non-competition with the food supply, and its capacity to release other value-added products during the biorefinery process. Biomass is rich in organic carbon that can be converted into biofuels, bio-based chemicals, and other value-added products. Different lignocellulose biomass sources, including agricultural activities, forest residues, algae, energy grass, municipal solid waste, and wastewater slurry, have been utilized for bioenergy production [4,5]. Lignocellulose biomass is a composition of primarily polar and non-polar constituents like lignin, hemicellulose, and cellulose that are strongly bound together, as illustrated in Figure 7.1. Biomass has a recalcitrant characteristic due to the cell wall arrangement, cellulose crystallinity, and high level of lignification, which makes it inaccessible during bioprocessing [6,7].

Accessing the simple sugars in cellulose during biorefinery is a major challenge because they are bound by hemicellulose and lignin through various bonds. Lignin is a major component of biomass that hinders the application of enzymes and biocatalysts to hydrolyze polymeric components, such as hemicellulose and cellulose, during the downstream processing of bioprocessing [9]. Therefore, the application of pretreatment is required to disrupt the biomass's morphological arrangement and fully access its polysaccharide content during biorefinery. Different pretreatment techniques, such as chemical, biological, mechanical, thermal, and combined methods, have been investigated [10]. Conventional pretreatment methods were noticed to have challenges like the need for a catalyst, washing after pretreatment, poor

DOI: 10.1201/9781003633815-10

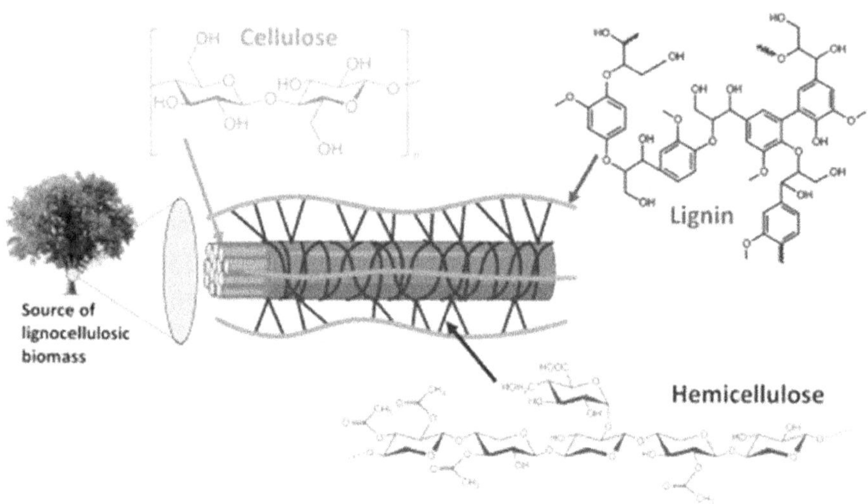

FIGURE 7.1 Composition of lignocellulose biomass [8].

stability, etc. These drawbacks limit the application of conventional pretreatment techniques and necessitate the need to find sustainable, economical, effective, and eco-benign pretreatment methods. To overcome these challenges, various methods, including the use of nanoparticles and combined pretreatment techniques, have received special attention. Nanomaterials are widely used in various applications, including pollutant removal, biomass conversion, medicine, biosensing, refrigeration, and biofuel generation, due to their numerous benefits, such as recyclability, high stability, good catalytic properties, and low production costs [11]. The application of nanoparticles in bioprocessing aims to improve process efficiency by enhancing heat and mass transfer, as well as both enzymatic and metabolic activities. It can be achieved using their large surface areas, enzyme cofactor activity, and catalytic properties [12]. Several studies have investigated the application of nanomaterials to enhance biorefinery processes, reporting improvements in enzymatic hydrolysis and biofuel production efficiency. Combined pretreatment is another important technique that has been investigated to overcome some of the challenges associated with conventional pretreatments. This involves the combination of two or more pretreatment techniques, where the methods complement one another [13]. This technique combines methods with varying limitations, ensuring that the limitations of each method are considered. Despite the advancement and successes recorded in the use of nanoparticles and combined pretreatment to improve the biorefinery process, there are still some limitations to the techniques. This chapter provides an overview of the application of nanoparticles and combined pretreatment techniques in bioenergy production, aiming to enhance understanding of their use. It highlights the fundamentals of each pretreatment technique, the progress made, and identifies some of the limitations of the methods. The areas of further study required to enhance the performance of the methods are also presented in this chapter. The social and environmental impacts of the methods are detailed in the chapter.

7.2 NANOPARTICLE ADDITIVES

Conventional biomass pretreatment techniques have some drawbacks that limit their ability to minimize the degree of polymerization and cellulose crystallinity. The generation of inhibitory materials, partial digestibility of feedstock, and the release of monomer compounds of lignin are some of the challenges for the bioenergy downstream process. The primary drawback is how to completely disrupt the lignin content, cost-efficiently, and still maintain the purity of sugars needed for subsequent biofuel production [14]. The introduction of nanotechnology has been identified as an alternative that can significantly influence biomass biorefineries due to its properties [15]. Nanotechnology is a rapidly emerging field of science that began in the 21st century, and it is believed to play a significant role in biorefineries [16]. This technology involves the process of synthesizing, manipulating, and utilizing materials with a size range of 1 to 100 nm. It is a building block of nanomaterials that has special and extraordinary properties. It has novel biological, chemical, and physical characteristics (such as high reactivity, minute size, high catalytic performance, and large surface area-to-volume ratio), which make its application in several fields like biomedicine, food industries, electronics, and agriculture [17]. Recently, various researchers have investigated the potential of nanomaterials in biorefineries, observing their potential as a pretreatment technique for biomass [18,19]. Nanomaterial additives in biorefineries can have a positive or negative influence on the downstream process of bioenergy, depending on the type of biomass, nanoparticle properties, biorefinery process, and process conditions [20]. The penetrating strength of nanoparticles enables them to penetrate the cell wall of the biomass, thereby increasing the surface area. The interaction of nanoparticles with other components of the biomass results in the release of lignin or hemicellulose under severe conditions [21]. Due to their considerable advantages in catalysis and enzyme immobilization, nanomaterials such as carbon-based nanoparticles (e.g., graphene and carbon nanotubes (CNTs)), and magnetic nanoparticles (MNPs) are commonly used in biofuel production. Metallic nanoparticles produced from nickel, cobalt, and platinum are most suitable for use during transesterification, which is crucial for biodiesel production [22]. Biomethane production from biomass has been observed to benefit from nanoparticles generated from metals like aluminum, iron, titanium, copper, cerium, zinc, and manganese [23–25]. The application of nanotechnology in processes such as anaerobic digestion, pyrolysis, gasification, and hydrogenation has been shown to enhance biofuel production [26–28].

For instance, the application of metal oxide nanoparticles during anaerobic digestion was reported to convert CO_2 to CH_4 during the methanogenesis process [29]. During the methanogenesis stage, zero-valent iron and the ferrous ion (Fe^{2+}) nanoparticles act as potential electron donors, leading to improved biogas yield [30]. These nanoparticles can also produce crucial microbial enzymes necessary for anaerobic digestion, which is essential for generating biogas. Fe^{2+} is an important component of coenzyme F_{420}; therefore, the addition of zero-valent iron nanoparticles improves the performance of the coenzyme [31]. Lower concentrations of trace metals and metal oxides like iron, nickel, zinc, cobalt, molybdenum, copper, etc. can play a crucial role in the stimulation and stability of anaerobic digestion of lignocellulose biomass during biogas generation [32]. These metal ions have the capacity to behave as enzymes and cofactors in several biogas production processes, leading to enrichment of biogas generation. Cobalt is needed by methanogenic archaea as a protein cofactor of vitamin B12 for methanol degradation, and nickel is required as a F_{430} coenzyme during biogas production [33]. Because of its ability to act as an oxidation-reduction agent, coenzyme, and terminal electron acceptor, Fe metal ions are important for optimum microbial growth and performance in anaerobic digester. Copper is needed as a coenzyme and supports biological electron transport, and zinc is needed to boost enzymatic performance, like methyl coenzyme M, and as a trans-esterification factor for the structural ions [34,35]. Different types of nanoparticles have been used to enhance the performance of biofuel refinery, and some of them are discussed below.

7.2.1 Metal Oxide Nanoparticles

Metal oxide nanoparticles are further classified as alkaline earth metal oxides (MgO, CaO, mixed metal oxides (TiO_2-ZnO, iron-tin, iron-cadmium), transition metal oxides (nano-sulfated TiO_2), and supported metal oxide (KF/CaO-MgO, CsNanoMgO-C) nanoparticles [36]. The synthesis technique of metal oxide nanoparticles is fundamental to their performance, and solution-stage techniques define the synthesized product. They are usually arranged using the sol-gel approach, where the reaction is terminated before gelation occurs, similar to precipitation approaches. This type of nanoparticle is widely used for its applications in catalysis, electronic materials, environmental remediation, and sensors. It has been utilized for the transformation of vegetable oil into biofuel. Co_3O_4, MoO_3, NiO, V_2O_5, KOH, and ZnO are some of the metal oxide nanoparticles that have been used as catalysts during the conversion of oil into organic liquid products [37]. It was noticed to have a support system with high catalytic performance and lower selectivity. The use of SiO_2 in biohydrogen generation through dark fermentation has been reported to improve hydrogen yield. Nanoparticles generated through the combination of SiO_2 and Fe_3O_4 were observed to exhibit higher catalytic performance and stability, resulting in a significant increase in biohydrogen yield [38]. The application of Fe nanoparticles was reported to significantly lower H_2S in biogas and improve methane release [39]. The introduction of Fe^{2+} or Fe^{3+} through Fe_3O_4 nanoparticles during anaerobic digestion can serve as a growth supplement for the microbes and enhance their activities [28]. The characteristics of Fe_3O_4 were observed to have magnetite and a small amount of goethite, and magnetite releases bioavailable ions (Fe^{3+} and Fe^{2+}), which are important nutrients for microorganisms' development [40]. It enhances hydrogenotrophic methanogenesis by releasing hydrogen or electrons during iron corrosion and facilitates methane release through CO2 consumption, as illustrated in Equations 7.1–7.3 [41].

$$CO_2 + 4Fe^0 + 8H^+ \rightarrow CH_4 + 4Fe^{2+} + 2H_2O, \tag{7.1}$$

$$Fe^0 + 2H_2O \rightarrow Fe^{2+} + H_2 + 2OH^-, \tag{7.2}$$

$$CO_2 + 4H_2 \rightarrow CH_4 + 2H_2O. \tag{7.3}$$

7.2.2 Magnetic nanoparticles

MNPs are characterized by a higher surface-to-volume ratio, immobilizing influence, and quantity characteristics, which make them useful in several applications [42]. Lipases and cellulases are usually utilized during biofuel production, and findings indicate that MNPs play a crucial role in the immobilization of these enzymes for biofuel production [43]. Enzymes used for immobilization can be reused to assist a matrix coated with specific nanomaterials, and this technique is suitable for lignocellulose biomass hydrolysis [44]. Alteration of different characteristics of immobilization enzymes can help improve the performance of enzymes for hydrolyzing lignocellulose biomass. The supermagnetic characteristics of MNPs can be utilized to separate and immobilize enzymes, thereby improving their reusability [45]. Studies have investigated the immobilization of cellulose on MNPs for the purpose of biomass hydrolysis [46]. Commercial-scale biodiesel production is encouraged by the reuse benefits of immobilized enzymes, which can be used consecutively about 10 times [47]. Additionally, a magnetic field can be used to easily recover MNPs. Nanoparticles of $CaSO_4/Fe_2O_3$-SiO_2, MgO/La_2O_3, $MgO/MgAL_2O_4$, Ni/ZnO, and sono-sulfated zirconia/MCM have been reported to improve the biofuel production [48], and it was observed that cellulose immobilization with MNPs will play a crucial role in biofuel production.

7.2.3 Metallic nanoparticles

Metallic nanoparticles are another type of nanoparticle that has been successfully applied to immobilize lignocellulose enzymes, such as cellulase. Very few studies have investigated the use of metallic nanoparticles for enhancing biofuel production effectiveness. It is known for its nanosize, which allows several enzymes, such as oxidoreductase, to bind to MNPs, subsequently enhancing the transfer of electrons [49]. It has been observed that metallic nanoparticles can be incorporated in a structured manner to enhance electrocatalytic activity and produce a biofuel cell with high loading capacity and a favorable electron transfer rate when utilized in layer-by-layer assembly with suitable enzymes and polymers [50]. Metallic nanoparticles, such as crystalline nickel, have been experimented with in various biofuel generation processes, including biohydrogen generation from lignocellulose biomass and biomass for syngas pretreatment [51]. The use of 40-nm silica nanoparticles for simultaneous scarification and fermentation process enhanced ethanol generation during biomass hydrolysis. It was reported that immobilization with silica nanoparticles released 1.6 times more glucose than cellulase, with a subsequent 2.1–2.3 times improvement in ethanol generated [52]. Nanoparticles of gold, platinum, and silver origins have also been reported as good materials for cellulase immobilization. The biohydrogen production rate was reported to increase by 46% when 5 nm gold nanoparticles were applied to artificial wastewater, while feedstock utilization capacity was enhanced by 56% [53]. The performance of metallic nanoparticles is mostly influenced by the concentration used and particle size [18].

7.2.4 Carbon nanotubes (CNTs)

CNTs have been identified as another enzyme immobilization nanomaterial due to their unique arrangement and capacity to provide extraordinary characteristics, such as mechanical strength, thermal, and electrical properties [54]. They are catalytic and conductive materials that can act as redox mediators during the redox reactions involved in the metabolic degradation of biomass. They assist in catalysis, as they react quickly with various organic and inorganic materials during biofuel production due to their stability in both strong and weak pH media [55]. Findings show that granular activated carbon enhances the development of methanogens and speeds up the activities of methanogenic bacteria at the beginning of anaerobic digestion. CNTs are made up of graphite sheets arranged in a cylindrical form. It can be a single-walled or multiwalled type, depending on the number of layers. Multiwalled CNTs are more suitable for cellulase immobilization due to their enhanced electronic characteristics, improved chemical and physical stability, ease of preparation, lower cost, and reduced toxicity compared to single-walled CNTs [56]. Multiwalled CNTs were reported to have a direct influence on methanogens' performance, irrespective of the occurrence of other mechanisms like direct transfer of electrons by interspecies. The two significant methods of enzyme immobilization using CNTs in industry are covalent and non-covalent immobilization. CNTs have been applied in various biofuel production processes and have been reported to improve the release of biofuel from biomass. The addition of CNTs in the form of biochar, graphite, and carbon cloth to an anaerobic sludge blanket digester was observed to affect microbial activity, availability, methane release, and syntrophic ethanol metabolism. Multiwalled CNTs were also noticed to improve butyrate methane generation from microbial enrichments [57]. Introduction of 100 mg/l of CNTs during biohydrogen production from lignocellulose biomass in a upflow anaerobic sludge blanket (UASB) digester improved the biohydrogen release rate to 5.55 l/l/d and the biohydrogen yield to 2.45 mol/mol glucose. The reaction start-up was reduced due to the addition of CNTs, and the performance was improved compared to the digester with activated carbon particles [58].

Studies have shown that nanomaterials can significantly enhance biofuel production from lignocellulose biomass. It has been observed that it enhances enzyme production, performance, and hydrolysis efficiency during biofuel generation [28,48,57]. Furthermore, the ability to recover and reuse nanomaterials, especially MNPs, can help reduce the amount of pretreatment required. Therefore, nanotechnology has huge potential to replace existing conventional pretreatment methods, and it holds promise for the future of biofuel production from biomass by increasing process efficiency, reducing environmental impact, and supporting a global approach to shift to sustainable energy sources [59]. However, there are some challenges that limit the application of nanoparticle treatment in biofuel production, which need to be addressed to access its full potential. The application of some metal oxide nanoparticles (e.g., TiO_2, ZnO, CuO, and Al_2O_3) and CNTs in algae biorefinery is challenging. It has been observed that some of these nanomaterials have potent antimicrobial characteristics that can impede the development of microalgae by releasing oxidative stress, inconsistent nutrient supply, and agglomeration [60]. Another major limitation of this method is the cost of synthesis. The process of nanoparticle synthesis requires sophisticated equipment and a significant investment, which may increase the cost of biorefinery. The level of clear information about their impacts on the molecular chemistry level is very low, and there is a need for more clarity on this. To fully realize the potential of this method, further studies are needed to address some of its drawbacks. There is a need for an assessment of the safety of nanoparticles, as they are observed to have certain impacts on the environment and human beings. Studies and continuous research are necessary to establish a comprehensive understanding that can further enhance the application in systematic ways. Furthermore, the synthesis of sustainable and green nanomaterials will have a significant impact on the cost of pretreatment, making it economically viable to replace existing conventional methods of synthesis and pretreatment. This will also reduce the possibility of environmental pollution during the nanoparticle preparation.

7.3 COMBINED PRETREATMENT

The application of single pretreatment methods, such as biological, physical, thermal, chemical, or nano-additive pretreatments, in biomass pretreatment has been observed to have some significant drawbacks. Some of these limitations include high energy requirements, low conversion efficiency, chemical requirements, the release of inhibitory compounds, and limited application [61]. Due to the challenges associated with single pretreatment, combined pretreatment has been identified as a crucial technique to enhance the efficiency of biomass conversion. This is the use of more than one pretreatment technique on a single biomass during processing [62]. Combining more than one technique to enhance pretreatment can be a viable option for identifying a suitable technique to improve biomass fractionalization. For example, it can be used to generate process streams that are optimized for hemicelluloses, while another stream is optimized for another product. It uses the synergetic action of multiple methods for more efficient disruption of biomass arrangement and exposes hemicellulose and cellulose. This will improve the degree of availability of cellulose and enhance the subsequent downstream bioprocessing of biomass. The merits of this technique include synergistic impact and broader application [63,64]. The application of various pretreatment techniques incurs additional costs, especially when the techniques are dissimilar; therefore, it is advisable to combine similar pretreatment techniques.

Generally, pretreatment typically begins with particle size reduction, which can be categorized as mechanical pretreatment, especially when it is thorough. Combined pretreatment can be mechanical-biological, mechanical-chemical, mechanical-thermal, mechanical-mechanical, biological-chemical, chemo-mechanical, or mechanical-nano additive-based, among others. Mechanical-nano-additive combined pretreatment mostly uses particle size reduction with different nanoparticles to improve the accessibility of the biomass. Particle size reduction, a mechanical pretreatment, increases the surface area of the biomass and creates more space for the attachment of nanoparticles. This has been observed to improve the biomethane released from lignocellulose feedstock. Particle size reduction combined with Fe_3O_4 was reported to improve the biomethane yield of *Arachis hypogea* shells. A

20 mg/l Fe3O4 improved biomethane yield by 80.59%, but it increased by 106.66% when combined with a 6-mm particle size [28]. The major challenge with this technique is the potential for rapid hydrolysis, which can lead to the overaccumulation of volatile fatty acids when a smaller particle is used. Biological-alkaline combined pretreatment can enhance the delignification of complex biomass, helping to reduce chemical concentration, temperature, and exposure time, thereby lowering the process cost. Significant loss of carbohydrate during pretreatment is the major limitation to this technique [65,66]. Oxidative-nanoparticle additives utilize the strength of oxidative pretreatment to delignify the biomass by breaking down the microstructural arrangement of the feedstock, thereby providing more space for nanoparticle attachment. Oxidative pretreatment combined with Fe_3O_4 on *Xyris capensis* was observed to increase biomethane production by 56–72% compared to single pretreatment, resulting in a 48–55% increase [21]. In a similar study, biological and oxidative pretreatments were combined. The decaying strength of white-rot fungi and the Fenton-based oxidation characteristics were utilized. This can be mimicked with other oxidizing reagents, such as H_2O_2, followed by biological pretreatment, which reduces exposure time and improves biomass delignification with minimal release of inhibitory compounds, resulting in enhanced sugar yields. This technique appears to be the brightest and most effective among biological-chemical combined pretreatments [67,68]. It has been observed that the effectiveness of combined pretreatment techniques involving biological pretreatment depends primarily on the type of biomass, the microbial strains/species, the culture conditions, and the order in which the techniques are utilized [65]. Different combined pretreatment techniques have been extensively investigated in recent years to establish economically viable, most effective, and general solutions. The most efficient and effective combined pretreatment method has not yet been established, and further improvements in the pretreatment method are still needed to overcome the drawbacks of optimizing the use of lignocellulose biomass and make biorefinery more accessible and profitable [69,70]. Further research can be focused on the following areas:

I. Develop more suitable and compatible pretreatment equipment and technology that widens the scope of combined pretreatment applications.
II. Investigating the capabilities of chemical reagent sources and combining them with other techniques.
III. Investigating the synergetic mechanism of multiple pretreatment techniques.
IV. Develop techniques and approaches for estimating the efficiency of pretreatment procedures.
V. Examining the biological reaction mechanism of the digestion microbes to pretreatment techniques.

7.4 ENVIRONMENTAL AND SOCIAL IMPACTS

The environmental and social impacts of combined pretreatment techniques vary depending on the type of feedstock and the specific techniques employed. The technique can offer enhanced green credentials by minimizing the total energy required and the use of toxic chemicals, compared to the conventional single pretreatment method. Some combined pretreatment techniques can minimize greenhouse gas emissions and environmental factors, such as human health effects and eutrophication; however, the need for high temperatures, toxic chemicals, a high volume of water for washing, and high energy consumption during pretreatment processes results in an environmental footprint [71]. A detailed life cycle assessment (LCA) is crucial for fully analyzing the specific environmental and social impacts of each combined pretreatment of biomass. LCA is a tool to determine the environmental impediment associated with all the processes involved in combined pretreatment. This is required to identify the best-combined pretreatment method from the environmental and social perspectives. Therefore, a detailed assessment of combined techniques is crucial to significantly reduce the environmental and social impacts of combined pretreatment during bioenergy production [72].

7.5 CONCLUSION

Studies have shown that the use of nanoparticles has the potential to enhance biomass biorefinery and increase the effectiveness of the technology. Available literature established that this technique could replace the existing conventional pretreatment techniques. The use of MNPs as biocatalysts is a novel process that can further improve the biorefinery of biomass. To make this technique more acceptable and practically viable, further studies are needed to understand the process mechanism of nanoparticles during the downstream biofuel production process. Another area of research to support nanoparticle pretreatment is the production of nanomaterials from biomass. This will reduce the cost of nanoparticles, making the technique economically viable.

Regarding the combined pretreatment methods, the literature review revealed that this process can help overcome some of the challenges associated with conventional pretreatment methods. Several methods have been combined, and it has been reported that this approach significantly improves biorefinery efficiency. Most of these methods are still in the pilot phase, and scalability remains a significant challenge. The operating cost for most of these combinations is high due to the different stages/equipment required, and there is little information available about the long-term environmental effects of the technique. Generally, combined pretreatment methods offer promising potential for enhancing biomass utilization and can make a significant contribution to achieving sustainable energy and eco-benign targets. Therefore, concerted efforts are necessary to enhance the efficiency of the technique, particularly in terms of cost and environmental impact.

REFERENCES

[1] S. Dutta, S. Saravanabhupathy, Anusha, R.C. Rajak, R. Banerjee, P.K. Dikshit, C.T. Padigala, A.K. Das, B.S. Kim, Recent developments in lignocellulosic biofuel production with nanotechnological intervention: An emphasis on ethanol, Catalysts 13 (2023) 1439. https://doi.org/10.3390/CATAL13111439.

[2] M. Rajamehala, R. Su, B. Gopalakrishnan, A. Muthu Kumara Pandian, M. Vijay Pradhap Singh, S. Chozhavendhan, Boundaries and openings of biorefineries towards sustainable biofuel production, Biofuels and Bioenergy: A Techno-Economic Approach (2022) 3–22. https://doi.org/10.1016/B978-0-323-90040-9.00019-9.
[3] S. Venkata Mohan, S. Dahiya, K. Amulya, R. Katakojwala, T.K. Vanitha, Can circular bioeconomy be fueled by waste biorefineries—A closer look, Bioresour Technol Rep 7 (2019) 100277. https://doi.org/10.1016/J.BITEB.2019.100277.
[4] Z.Z. Rasmeni, D.M. Madyira, A.N. Matheri, O.O. Olatunji, K.O. Olatunji, Mechanical pre-treatment of co-digested Wastewater Sludge (WWS) and Brewery Spent Yeast (BSY) under anaerobic condition, in: International Conference on Electrical and Computer Engineering Researches, ICECER 2024 (2024). https://doi.org/10.1109/ICECER62944.2024.10920381.
[5] A.D. Olugbemide, B. Likozar, Assessment of liquid and solid digestates from anaerobic digestion of rice husk as potential biofertilizer and nutrient source for microalgae cultivation, Processes 10 (2022) 1007. https://doi.org/10.3390/PR10051007/S1.
[6] Z. Usmani, M. Sharma, P. Gupta, Y. Karpichev, N. Gathergood, R. Bhat, V.K. Gupta, Ionic liquid based pretreatment of lignocellulosic biomass for enhanced bioconversion, Bioresour Technol 304 (2020) 123003. https://doi.org/10.1016/J.BIORTECH.2020.123003.
[7] K.O. Olatunji, D.M. Madyira, Enhancement of methane yield of Arachis hypogea shells using chemical pretreatment techniques, Mater Today Proc 105 (2024) 189–193. https://doi.org/10.1016/J.MATPR.2023.08.284.
[8] R. Philippini, A.P. Ingle, S. Prasad, Emerging applications of nanomaterials in the pretreatment of lignocellulosic biomass, Nanotechnology for Biorefinery (2023) 117–141. https://doi.org/10.1016/B978-0-323-95965-0.00011-1.
[9] K.O. Olatunji, D.M. Madyira, Comparative analysis of the effects of five retreatment methods on morphological and methane yield of groundnut shells, Waste Biomass Valor 15 (2024) 469–486. https://doi.org/10.1007/S12649-023-02177-6/FIGURES/5.
[10] K.O. Olatunji, D.M. Madyira, J.O. Amos, Sustainable enhancement of biogas and methane yield of macroalgae biomass using different pretreatment techniques: A mini-review, Energy Environ 35 (2024) 1050–1088. https://doi.org/10.1177/0958305X231193869.
[11] Y. Zhou, B. Liu, R. Yang, J. Liu, Filling in the gaps between nanozymes and enzymes: Challenges and opportunities, Bioconjug Chem 28 (2017) 2903–2909. https://doi.org/10.1021/ACS.BIOCONJCHEM.7B00673/ASSET/IMAGES/MEDIUM/BC-2017-00673V_0005.GIF.
[12] S. Bosu, N. Rajamohan, Nanotechnology approach for enhancement in biohydrogen production—review on applications of nanocatalyst and life cycle assessment, Fuel 323 (2022) 124351. https://doi.org/10.1016/J.FUEL.2022.124351.
[13] K.O. Olatunji, N.A. Ahmed, O. Ogunkunle, Optimization of biogas yield from lignocellulosic materials with different pretreatment methods: A review, Biotechnol Biofuels 14 (2021) 1–34. https://doi.org/10.1186/S13068-021-02012-X.
[14] M. Yadav, A. Singh, V. Balan, N. Pareek, V. Vivekanand, Biological treatment of lignocellulosic biomass by Chaetomium globosporum: Process derivation and improved biogas production, Int J Biol Macromol 128 (2019) 176–183. https://doi.org/10.1016/J.IJBIOMAC.2019.01.118.
[15] I. Arya, A. Poona, P.K. Dikshit, S. Pandit, J. Kumar, H.N. Singh, N.K. Jha, H.A. Rudayni, A.A. Chaudhary, S. Kumar, Current trends and future prospects of nanotechnology in biofuel production, Catalysts 11 (2021) 1308. https://doi.org/10.3390/CATAL11111308.
[16] Y. Kumar, P. Yogeshwar, S. Bajpai, P. Jaiswal, S. Yadav, D.P. Pathak, M. Sonker, S.K. Tiwary, Nanomaterials: Stimulants for biofuels and renewables, yield and energy optimization, Mater Adv 2 (2021) 5318–5343. https://doi.org/10.1039/D1MA00538C.

[17] A.P. Ingle, M.P. Moharil, Role of nanotechnology in the biorefineries: Recent advances and challenges, Nanotechnology for Biorefinery (2023) 89–115. https://doi.org/10.1016/B978-0-323-95965-0.00012-3.
[18] K.O. Olatunji, D.M. Madyira, Z.Z. Rasmeni, Enhancing biogas yield of Xyris capensis grass using different nanoparticles additive, IOP Conf Ser Earth Environ Sci 1322 (2024) 012001. https://doi.org/10.1088/1755-1315/1322/1/012001.
[19] T.A.M. Abdelwahab, M.K. Mohanty, P.K. Sahoo, D. Behera, Metal nanoparticle mixtures to improve the biogas yield of cattle manure, Biomass Convers Biorefin 13 (2023) 2243–2254. https://doi.org/10.1007/S13399-021-01286-3/METRICS.
[20] C.M. Ajay, S. Mohan, P. Dinesha, M.A. Rosen, Review of impact of nanoparticle additives on anaerobic digestion and methane generation, Fuel 277 (2020) 118234. https://doi.org/10.1016/J.FUEL.2020.118234.
[21] K.O. Olatunji, D.M. Madyira, Enhancing microstructural and biomethane yield of Xyris capensis using combined oxidative and nanoparticle additive pretreatment methods, Biomass Convers Biorefin 1 (2023) 1–15. https://doi.org/10.1007/S13399-023-04401-8/FIGURES/5.
[22] N. Singh, B.S. Dhanya, M.L. Verma, Nano-immobilized biocatalysts and their potential biotechnological applications in bioenergy production, Mater Sci Energy Technol 3 (2020) 808–824. https://doi.org/10.1016/J.MSET.2020.09.006.
[23] K.O. Olatunji, D.M. Madyira, N.A. Ahmed, O. Ogunkunle, Experimental evaluation of the influence of combined particle size pretreatment and Fe<inf>3</inf>O<inf>4</inf> additive on fuel yields of Arachis Hypogea shells, Waste Manag Res 41 (2023). https://doi.org/10.1177/0734242X221122560.
[24] H. Baniamerian, P.G. Isfahani, P. Tsapekos, M. Alvarado-Morales, M. Shahrokhi, M. Vossoughi, I. Angelidaki, Application of nano-structured materials in anaerobic digestion: Current status and perspectives, Chemosphere 229 (2019) 188–199. https://doi.org/10.1016/J.CHEMOSPHERE.2019.04.193.
[25] E. Mulu, M.M. M'Arimi, R.C. Ramkat, A review of recent developments in application of low cost natural materials in purification and upgrade of biogas, Renew Sustain Energy Rev 145 (2021) 111081. https://doi.org/10.1016/J.RSER.2021.111081.
[26] S. Ali, O. Shafique, S. Mahmood, T. Mahmood, B.A. Khan, I. Ahmad, Biofuels production from weed biomass using nanocatalyst technology, Biomass Bioenergy 139 (2020) 105595. https://doi.org/10.1016/J.BIOMBIOE.2020.105595.
[27] S.T. Hussain, S.A. Ali, A. Bano, T. Mahmood, Use of nanotechnology for the production of biofuels from butchery waste, Int J Phys Sci 6 (2011) 7271–7279. https://doi.org/10.5897/IJPS11.992.
[28] K.O. Olatunji, D.M. Madyira, N.A. Ahmed, O. Ogunkunle, Effect of combined particle size reduction and Fe<inf>3</inf>O<inf>4</inf> additives on biogas and methane yields of Arachis hypogea shells at mesophilic temperature, Energies (Basel) 15 (2022). https://doi.org/10.3390/en15113983.
[29] B.S. Zakaria, B.R. Dhar, Progress towards catalyzing electro-methanogenesis in anaerobic digestion process: Fundamentals, process optimization, design and scale-up considerations, Bioresour Technol 289 (2019) 121738. https://doi.org/10.1016/J.BIORTECH.2019.121738.
[30] E. Abdelsalam, M. Samer, Y.A. Attia, M.A. Abdel-Hadi, H.E. Hassan, Y. Badr, Influence of zero valent iron nanoparticles and magnetic iron oxide nanoparticles on biogas and methane production from anaerobic digestion of manure, Energy 120 (2017) 842–853. https://doi.org/10.1016/J.ENERGY.2016.11.137.
[31] R. Barrena, M. del C. Vargas-García, G. Capell, M. Barańska, V. Puntes, J. Moral-Vico, A. Sánchez, X. Font, Sustained effect of zero-valent iron nanoparticles under semi-continuous anaerobic digestion of sewage sludge: Evolution of nanoparticles

and microbial community dynamics, Sci Total Environ 777 (2021) 145969. https://doi.org/10.1016/J.SCITOTENV.2021.145969.

[32] S. Zhang, Y. Ren, X. Ma, W. Guan, M. Gao, Y.Y. Li, Q. Wang, C. Wu, Effect of zero-valent iron addition on the biogas fermentation of food waste after anaerobic preservation, J Environ Chem Eng 9 (2021) 106013. https://doi.org/10.1016/J.JECE.2021.106013.

[33] F.R. Amin, H. Khalid, H.M. El-Mashad, C. Chen, G. Liu, R. Zhang, Functions of bacteria and archaea participating in the bioconversion of organic waste for methane production, Sci Total Environ 763 (2021) 143007. https://doi.org/10.1016/J.SCITOTENV.2020.143007.

[34] Y.C. Chou, J.J. Su, Biogas production by anaerobic co-digestion of dairy wastewater with the crude glycerol from slaughterhouse sludge cake transesterification, Animals 9 (2019) 618. https://doi.org/10.3390/ANI9090618.

[35] Y. Li, J. Zhao, J. Krooneman, G.J.W. Euverink, Strategies to boost anaerobic digestion performance of cow manure: Laboratory achievements and their full-scale application potential, Sci Total Environ 755 (2021) 142940. https://doi.org/10.1016/J.SCITOTENV.2020.142940.

[36] Y. Kumar, P. Yogeshwar, S. Bajpai, P. Jaiswal, S. Yadav, D.P. Pathak, M. Sonker, S.K. Tiwary, Nanomaterials: Stimulants for biofuels and renewables, yield and energy optimization, Mater Adv 2 (2021) 5318–5343. https://doi.org/10.1039/D1MA00538C.

[37] Z.D. Yigezu, K. Muthukumar, Catalytic cracking of vegetable oil with metal oxides for biofuel production, Energy Convers Manag 84 (2014) 326–333. https://doi.org/10.1016/J.ENCONMAN.2014.03.084.

[38] S.V. Mohan, G. Mohanakrishna, S.S. Reddy, B.D. Raju, K.S.R. Rao, P.N. Sarma, Self-immobilization of acidogenic mixed consortia on mesoporous material (SBA-15) and activated carbon to enhance fermentative hydrogen production, Int J Hydrogen Energy 33 (2008) 6133–6142. https://doi.org/10.1016/J.IJHYDENE.2008.07.096.

[39] L. Su, X. Shi, G. Guo, A. Zhao, Y. Zhao, Stabilization of sewage sludge in the presence of nanoscale zero-valent iron (nZVI): Abatement of odor and improvement of biogas production, J Mater Cycles Waste Manag 15 (2013) 461–468. https://doi.org/10.1007/s10163-013-0150-9.

[40] A. El Nemr, M.A. Hassaan, M.R. Elkatory, S. Ragab, A. Pantaleo, Efficiency of Fe3O4 nanoparticles with different pretreatments for enhancing biogas yield of macroalgae ulva intestinalis linnaeus, Molecules 26 (2021) 5105. https://doi.org/10.3390/MOLECULES26165105.

[41] R. Xu, S. Xu, L. Zhang, A.P. Florentino, Z. Yang, Y. Liu, Impact of zero valent iron on blackwater anaerobic digestion, Bioresour Technol 285 (2019). https://doi.org/10.1016/J.BIORTECH.2019.121351.

[42] M. Ahmed, M. Douek, The role of magnetic nanoparticles in the localization and treatment of breast cancer, Biomed Res Int 2013 (2013). https://doi.org/10.1155/2013/281230.

[43] D.T. Tran, C.L. Chen, J.S. Chang, Immobilization of Burkholderia sp. lipase on a ferric silica nanocomposite for biodiesel production, J Biotechnol 158 (2012) 112–119. https://doi.org/10.1016/J.JBIOTEC.2012.01.018.

[44] M. Puri, C.J. Barrow, M.L. Verma, Enzyme immobilization on nanomaterials for biofuel production, Trends Biotechnol 31 (2013) 215–216. https://doi.org/10.1016/J.TIBTECH.2013.01.002.

[45] J. Alftrén, T.J. Hobley, Covalent immobilization of β-Glucosidase on Magnetic particles for Lignocellulose Hydrolysis, Appl Biochem Biotechnol 169 (2013) 2076–2087. https://doi.org/10.1007/S12010-013-0122-5/FIGURES/6.

[46] P.J. Huang, K.L. Chang, J.F. Hsieh, S.T. Chen, Catalysis of rice straw hydrolysis by the combination of immobilized cellulase from aspergillus niger on β-Cyclodextrin-Fe3O4

nanoparticles and ionic liquid, Biomed Res Int 2015 (2015) 409103. https://doi.org/10.1155/2015/409103.
[47] S. Duraiarasan, S.A. Razack, A. Manickam, A. Munusamy, M.B. Syed, M.Y. Ali, G.M. Ahmed, M.S. Mohiuddin, Direct conversion of lipids from marine microalga C. salina to biodiesel with immobilised enzymes using magnetic nanoparticle, J Environ Chem Eng 4 (2016) 1393–1398. https://doi.org/10.1016/J.JECE.2015.12.030.
[48] S.H. Teo, A. Islam, E.S. Chan, S.Y. Thomas Choong, N.H. Alharthi, Y.H. Taufiq-Yap, M.R. Awual, Efficient biodiesel production from Jatropha curcus using CaSO4/Fe2O3-SiO2 core-shell magnetic nanoparticles, J Clean Prod 208 (2019) 816–826. https://doi.org/10.1016/J.JCLEPRO.2018.10.107.
[49] K.A. Vincent, X. Li, C.F. Blanford, N.A. Belsey, J.H. Weiner, F.A. Armstrong, Enzymatic catalysis on conducting graphite particles, Nat Chem Biol3 (2007) 761–762. https://doi.org/10.1038/nchembio.2007.47.
[50] C.H. Kwon, Y. Ko, D. Shin, M. Kwon, J. Park, W.K. Bae, S.W. Lee, J. Cho, High-power hybrid biofuel cells using layer-by-layer assembled glucose oxidase-coated metallic cotton fibers, Nat Commun 9 (2018). https://doi.org/10.1038/S41467-018-06994-5.
[51] Y. Richardson, J. Blin, G. Volle, J. Motuzas, A. Julbe, In situ generation of Ni metal nanoparticles as catalyst for H2-rich syngas production from biomass gasification, Appl Catal A Gen 382 (2010) 220–230. https://doi.org/10.1016/J.APCATA.2010.04.047.
[52] J.S. Lupoi, E.A. Smith, Evaluation of nanoparticle-immobilized cellulase for improved ethanol yield in simultaneous saccharification and fermentation reactions, Biotechnol Bioeng 108 (2011) 2835–2843. https://doi.org/10.1002/BIT.23246.
[53] Y. Zhang, J. Shen, Enhancement effect of gold nanoparticles on biohydrogen production from artificial wastewater, Int J Hydrogen Energy 32 (2007) 17–23. https://doi.org/10.1016/J.IJHYDENE.2006.06.004.
[54] R. Ahmad, S.K. Khare, Immobilization of Aspergillus niger cellulase on multiwall carbon nanotubes for cellulose hydrolysis, Bioresour Technol 252 (2018) 72–75. https://doi.org/10.1016/J.BIORTECH.2017.12.082.
[55] G. Sierra Gallego, J. Barrault, C. Batiot-Dupeyrat, F. Mondragón, Production of hydrogen and MWCNTs by methane decomposition over catalysts originated from LaNiO3 perovskite, Catal Today 149 (2010) 365–371. https://doi.org/10.1016/J.CATTOD.2009.06.004.
[56] J.H. Lehman, M. Terrones, E. Mansfield, K.E. Hurst, V. Meunier, Evaluating the characteristics of multiwall carbon nanotubes, Carbon N Y 49 (2011) 2581–2602. https://doi.org/10.1016/J.CARBON.2011.03.028.
[57] C.E. Kozonoe, R. de Paiva Floro Bonfim, R.M. Brito Alves, M. Schmal, The Fe-Co-Cu supported on MWCNT as catalyst for the tri-reforming of methane—Investigating the structure changes of the catalysts, Fuel 256 (2019) 115917. https://doi.org/10.1016/J.FUEL.2019.115917.
[58] Z. Liu, F. Lv, H. Zheng, C. Zhang, F. Wei, X.H. Xing, Enhanced hydrogen production in a UASB reactor by retaining microbial consortium onto carbon nanotubes (CNTs), Int J Hydrogen Energy 37 (2012) 10619–10626. https://doi.org/10.1016/J.IJHYDENE.2012.04.057.
[59] D. Rathi, R.K. Kapoor, P. Singh, P. Nandal, Applications of nanotechnology in the production of biofuels: Emphasis on nanomaterials and lignocellulosic biomass pretreatment, Lignocellulosic Biomass and Enzymes (2025) 271–294. https://doi.org/10.1007/978-981-96-3037-0_12.
[60] K.S. Khoo, W.Y. Chia, D.Y.Y. Tang, P.L. Show, K.W. Chew, W.H. Chen, Nanomaterials utilization in biomass for biofuel and bioenergy production, Energies 13 (2020) 892. https://doi.org/10.3390/EN13040892.

[61] S. Ma, Y. Li, J. Li, X. Yu, Z. Cui, X. Yuan, W. Zhu, H. Wang, Features of single and combined technologies for lignocellulose pretreatment to enhance biomethane production, Renew Sustain Energy Rev 165 (2022) 112606. https://doi.org/10.1016/J.RSER.2022.112606.
[62] K. Zeng, X. He, H. Yang, X. Wang, H. Chen, The effect of combined pretreatments on the pyrolysis of corn stalk, Bioresour Technol 281 (2019) 309–317. https://doi.org/10.1016/J.BIORTECH.2019.02.107.
[63] J.C. Martínez-Patiño, T.A. Lu-Chau, B. Gullón, E. Ruiz, I. Romero, E. Castro, J.M. Lema, Application of a combined fungal and diluted acid pretreatment on olive tree biomass, Ind Crops Prod 121 (2018) 10–17. https://doi.org/10.1016/J.INDCROP.2018.04.078.
[64] C. Areepak, T. Jiradechakorn, S. Chuetor, C. Phalakornkule, M. Sriariyanun, M. Raita, V. Champreda, N. Laosiripojana, Improvement of lignocellulosic pretreatment efficiency by combined chemo—Mechanical pretreatment for energy consumption reduction and biofuel production, Renew Energy 182 (2022) 1094–1102. https://doi.org/10.1016/J.RENENE.2021.11.002.
[65] S. Meenakshisundaram, A. Fayeulle, E. Leonard, C. Ceballos, A. Pauss, Fiber degradation and carbohydrate production by combined biological and chemical/physicochemical pretreatment methods of lignocellulosic biomass—A review, Bioresour Technol 331 (2021) 125053. https://doi.org/10.1016/J.BIORTECH.2021.125053.
[66] M. Si, D. Liu, M. Liu, X. Yan, C. Gao, L. Chai, Y. Shi, Complementary effect of combined bacterial-chemical pretreatment to promote enzymatic digestibility of lignocellulose biomass, Bioresour Technol 272 (2019) 275–280. https://doi.org/10.1016/J.BIORTECH.2018.10.036.
[67] S.R. Paudel, S.P. Banjara, O.K. Choi, K.Y. Park, Y.M. Kim, J.W. Lee, Pretreatment of agricultural biomass for anaerobic digestion: Current state and challenges, Bioresour Technol 245 (2017) 1194–1205. https://doi.org/10.1016/j.biortech.2017.08.182.
[68] C. Xie, W. Gong, Q. Yang, Z. Zhu, L. Yan, Z. Hu, Y. Peng, White-rot fungi pretreatment combined with alkaline/oxidative pretreatment to improve enzymatic saccharification of industrial hemp, Bioresour Technol 243 (2017) 188–195. https://doi.org/10.1016/J.BIORTECH.2017.06.077.
[69] B.O. Abo, M. Gao, Y. Wang, C. Wu, H. Ma, Q. Wang, Lignocellulosic biomass for bioethanol: An overview on pretreatment, hydrolysis and fermentation processes, Rev Environ Health 34 (2019) 57–68. https://doi.org/10.1515/REVEH-2018-0054.
[70] S. Rezania, B. Oryani, J. Cho, A. Talaiekhozani, F. Sabbagh, B. Hashemi, P.F. Rupani, A.A. Mohammadi, Different pretreatment technologies of lignocellulosic biomass for bioethanol production: An overview, Energy 199 (2020) 117457. https://doi.org/10.1016/J.ENERGY.2020.117457.
[71] A. Ajayi-Banji, G. Pourhashem, S. Rahman, X. Feng, Evaluating the environmental impacts of pretreatment and nanoparticles in solid-state anaerobic digestion using life cycle assessment, Bioenergy Res 17 (2024) 1971–1984. https://doi.org/10.1007/S12155-024-10757-0/FIGURES/4.
[72] S. Balchandani, U.R. Gracida-Alvarez, T.R. Hawkins, P.T. Benavides, Greenhouse gas emissions associated with pretreatment techniques utilized for biosugar production, ACS Sustain Chem Eng 13 (2025) 7469–7480. https://doi.org/10.1021/ACSSUSCHEMENG.5C00873/ASSET/IMAGES/LARGE/SC5C00873_0003.JPEG.

8 Techno-economic analysis of pretreatment methods

8.1 INTRODUCTION

A techno-economic analysis of pretreatment is necessary to determine the feasibility and profitability of a pretreatment method. It has been identified as an important tool for understanding potential feasibility, cost benchmarks, required studies, and the development of areas that may be needed, as well as decision-making that will be useful for upscaling and commercialization [1]. A techno-economic analysis is required to evaluate capital investment, manufacturing costs, operating costs, and profitability indicators. This process is a fundamental tool that assists researchers in developing new technology, integrating process modeling, and engineering design with economic analysis [2]. It can be utilized to identify bottlenecks and crucial aspects in the pretreatment process, thereby enhancing its technical or economic feasibility [3]. There are various techno-economic models for the pretreatment process, which can be used to evaluate the economic impact of production parameters [4]. Real economic figures can be utilized to validate the results from these models. This chapter provides an overview of the different components of techno-economic analysis that cumulate into decision-making about the sustainability of various pretreatment technologies. Different case scenarios were also considered to demonstrate the importance of techno-economic analysis and its significance in the selection of the pretreatment method. The chapter provides baseline information that can assist decision-makers, researchers, and industrialists in selecting the appropriate pretreatment.

8.2 PROCESS SYNTHESIS AND DESIGN

The design of a sustainable pretreatment process is a primary challenge for developing an eco-friendly and profitable biorefinery industry. This can be achieved through the integration of environmental, social, and economic criteria, using the life cycle assessment (LCA) strategies. This approach is necessary in biomass pretreatment, as the process aims to produce eco-friendly fuels and other value-added products that have the potential to facilitate the transition to a sustainable environment [5]. The proper starting point for biomass pretreatment design and synthesis is understanding the potential technique inputs and considering the main product lines possible from different applied treatment techniques. The origin and type of biomass feedstock (i.e., algae, agricultural residues, energy crops, municipal organic waste, or industrial waste), as well as the morphological arrangement of the biomass, significantly

DOI: 10.1201/9781003633815-11

influence the pretreatment applied and the subsequent fuel and chemicals produced [6]. For instance, biomass with a very high lignin percentage requires harsh pretreatment techniques (e.g., thermal, chemical with high concentrations of chemicals), which could result in the release of inhibitory compounds [7]. Therefore, an initial empirical analysis of the lignocellulosic properties of the intended biomass (heterogeneous or homogeneous) is required [8]. Further studies on the application of pretreatment to selected biomass components to produce important intermediates or final products in the value chain are needed to be investigated either experimentally or through previous studies on the selected biomass. One of the challenges of the existing studies is that they focused on a particular biomass cultivated/recovered under specific conditions. This becomes a challenge when the proposed biorefinery process plan involves different biomass feedstock and multiple conversion technologies for the purpose of generating multiple bioproducts [9]. Even when the potential of individual pretreatment methods has been established, the limitation in biorefinery design is the selection and allocation of optimal conditions for pretreatment and the generation of the intended product from biomass.

Different methods have been examined to assist in the design and synthesis of biomass pretreatment in biorefinery. Various process synthesis techniques, such as symbol triangle, Gibbs free energy, matrix synthesis, and retro-synthesis, have been investigated. However, the rise of environmental challenges has necessitated the integration of environmental considerations in the synthesis of biorefinery [10]. A systematic technique has been investigated for designing optimal pretreatment conditions in a biorefinery. This technique relied on advancing a superstructure of pretreatment methods and products and applying a tree-branching and searching method to select the most suitable method for the specified biomass and biorefinery process [11]. With several synthesis routes available, this method offers a rapid screening technique to minimize the number of pretreatment methods and identify a suitable one for degrading the selected biomass during the biorefinery process [12]. A combination of an upscale laboratory-scale study with a revision of the process configurations in line with the observed feasibility of the proposed commercial-scale pretreatment is another means of pretreatment process design. This method is time-consuming, but it is beneficial, especially for selecting bright pretreatment methods without any retrofitting challenges [8]. The optimization design method is one of the most popular methods for designing pretreatment applications. This method aims to maximize either energy consumption, mass yield, or economic potential, considering various challenges related to the selected method, reactants, conversion, and products [13]. This method helps minimize waste streams to improve sustainability. The first stage of this technique involves designing a pretreatment method by formulating an optimization problem (e.g., optimizing the profit margin or reducing process costs) [14]. It requires the formulation of a set of objectives based on the limitations and expected objectives. This takes place after identifying all the inputs involved in the pretreatment. Optimization software, such as response surface methodology (RSM), artificial neural networks (ANN), and adaptive neuro-fuzzy inference systems (ANFIS), has been utilized to optimize pretreatment methods [15,16]. The primary aim of the method is to determine the most efficient pretreatment by simultaneously evaluating varying input parameters. In general, the method aims to determine the most suitable pretreatment parameters through detailed integration and evaluation of various techniques under different conditions. Additionally, the optimization method aims to unravel the basis of the techno-economic analysis by evaluating the objective function [17].

Conceptual design, also known as a knowledge-based approach, is another important pretreatment design technique. This method employs a holistic approach, encompassing all areas related to biomass pretreatment from an engineering perspective, as well as the incorporation of social factors, to develop a pretreatment method that enhances the downstream biorefinery process. It is similar to the method used in recent years for the conventional conceptual design technique of chemical equipment [18]. The first stage in a pretreatment design involves selecting biomass and product based on their microstructural arrangements, proximate composition, ultimate properties, availability, season, and demand specifications. The production of other value-added products, such as lignin, enzymes, biomaterials, nutraceuticals, and pharmaceuticals, must be critically considered to make the process more economical [19]. This design approach encompasses three categories of analysis: technical, economic, and environmental, with two categories of integration: mass and energy [20]. The primary benefit of this method can be attributed to the understanding of purely conceptual approaches. It can incorporate an element of social analysis that is essential to understanding the concept of sustainability. Compared to the optimization method, this method does not require the formulation of a superstructure, as the approach and process configuration are selected based on hierarchy, sequencing, and integration strategies [21]. Biomass is a first-level hierarchy concept established during pretreatment, and different pretreatments can be considered depending on the type of biomass being considered. Therefore, a list of pretreatment methods and the corresponding expected biofuels can be predetermined based on the composition of the biomass. Several studies have investigated the pretreatment synthesis of biomass using either single-objective or multi-objective optimization, considering environmental effects, economic factors, risks, and other relevant aspects. However, a major limitation is the development of a tool that can be used to select the most suitable pretreatment approach. In fact, arbitrating and selecting between various pretreatment approaches can be complex. Therefore, there is a need for a novel approach that combines pretreatment design and optimization together with process assessment and a sustainability gauge, utilizing technical analysis and a weighted performance metric.

8.3 MASS BALANCE

Mass balance measures the inputs and outputs during the pretreatment process to ascertain the economic and environmental feasibility of the process. It is crucial for designing, optimizing, and corroborating the sustainability of the pretreatment process, ensuring its scalability and that the expected biofuels can be produced efficiently and economically [22]. It requires insights into the performance of pretreatment methods in terms of materials and greenhouse gas emissions during the process. This process considers the mass of all biomass components, including lignin, cellulose, hemicellulose, and ash, as they undergo the pretreatment process to ensure that the weight of biomass inputs is equivalent to the sum of the weights at the outputs [23]. It is essential for assessing pretreatment efficiency to substantiate the recovery of valuable materials and to verify material losses that can impact the economic feasibility of the process. Mass balance analysis allows the researcher to compare various pretreatment techniques by assessing the effectiveness of disrupting biomass into important biofuels [24]. The primary need of the mass-balance process is that all biomass and emissions from pretreatment are accounted for in the process. In short, it is the analysis of input biomass and output feedstock, with their emissions produced during the process [25]. To promote a particular pretreatment, it is crucial to use methods that have high biomass output after pretreatment and low emissions. Apart from the mass of biomass input to output considerations, a good pretreatment must tend toward net-zero emissions [26].

Mass balances for the conversion of biomass during the pretreatment process and the downstream process are crucial to ascertain the most suitable pretreatment methods [27]. Several pretreatment methods produce a liquid stream, and the quantity and type of components that are solubilized depend on the method used. Therefore, if the process yields are estimated according to the initial biomass weight without considering any pretreatment mass losses resulting from solid-liquid separation or post-washing, the estimation will be inaccurate.

Carbon balance is one of the mass balances that can be used to track the carbon-based components in the entire process. Carbon balance is difficult to ascertain accurately due to the complexity of the carbon-based biomass [28]. Another technique that has been used for mass balance is the estimation of different components of interest, like xylose/xylan, glucan/glucose, lignin, and other sugar components [29]. Mass balance can also be calculated using cumulative biomass components, using the total and volatile solids of the biomass [30]. However, the method is less reliable due to the difficulty involved in estimating the solid content of dilute liquids or solution streams.

8.4 NET ENERGY

Net energy analysis is used to determine the cumulative energy input, energy generated, and energy balance during the biorefinery process. This process is necessary to understand the balance between energy consumed and energy released, thereby determining the economic feasibility of the pretreatment method [31]. Most studies used life-cycle energy analysis to assess the energy yield of biofuels and calculate the energy used for pretreatment as a component of the energy consumed [32]. Energy is gained when the surplus energy, after the energy consumed, is deducted from the gross energy released during the biorefinery process. Since energy used during pretreatment is part of the energy required for the production process, the pretreatment method plays a significant role in determining the net energy. The efficiency of pretreatment on biofuel yield and the amount of energy required determine the net energy. Therefore, it is essential to select an efficient pretreatment with lower energy requirements to achieve a significant net energy gain [33]. Net energy can be calculated as illustrated in Equation 8.1 [31].

$$Net\,energy = energy\,generated - energy\,consumed. \tag{8.1}$$

The net energy result enables the quantification of the energy potential of different biomass feedstocks through pretreatment. It indicates whether the use of a pretreatment process for biofuel production enhances the potential of the biomass during the biorefinery process [34]. In addition, combining net energy analysis with the life-cycle assessment and techno-economic analysis provides a multicriteria analysis tool for biomass pretreatment that can be utilized to understand the trade-off between energy content, costs, and environmental impact [35,36]. Mass balance is also important in determining the energy balance since it is required to calculate the mass of the biomass generated and the energy it releases. These two parameters are needed to provide detailed information on a biorefinery's performance and sustainability. For instance, mass balance information can be used during a LCA to trace environmental flows, including energy inputs and outputs, and determine the total influence of the process.

8.5 COST OF EQUIPMENT AND CONSUMABLES

Cost analysis is an integral part of the design and feasibility study of the pretreatment process. A detailed economic analysis should encompass capital investment, operating cost analysis, and profitability analysis. The total cost of the pretreatment technique can be determined by summing the costs of its various components [37]. Biomass, equipment, energy, and chemicals/enzymes are some of the key cost components in the pretreatment process. The cost of pretreatment heavily depends on the type of biomass, the pretreatment method, and the investment cost for equipment, which also varies by type. Studies have shown that chemical pretreatment requires high investment costs due to the specialized equipment needed, which must have high resistance to corrosion. Mechanical pretreatment also required equipment and energy, necessitating high capital costs, whereas thermal pretreatment required a substantial investment [38–39]. Another important cost component of pretreatment methods is the cost of consumables, which varies depending on the type of pretreatment. These consumables include biological (e.g., enzymes, fungi, bacteria, etc.), chemicals (alkalis, acids, oxidizing agents, ionic solvents, ammonia, water, etc.), nanoparticles, and thermal (e.g., energy, water, etc.). These costs are significant to the process scale and specific parameters, such as solid–liquid ratio, solvent recovery, and energy efficiency [40,41]. The overall economic feasibility of pretreatment depends on the handling of biomass, waste treatment, and the downstream process. It has been observed that the cost estimate for the whole process at the laboratory scale is calculated based on the biomass used, the technique applied, and the energy required. But, at the pilot or commercial scale, cost calculation for various financial attributes is also crucial [42]. Therefore, a pretreatment technique with a low-cost attribute is crucial to the profitability of the overall biomass biorefinery process. Some pretreatment techniques are still at the laboratory scale; therefore, their profitability at the industrial level can be estimated using a shortcut for a preliminary assessment, utilizing correlations and reference investment costs from similar techniques. Net present value (NPV), discounted cash flow rate (DCFR), return on investment (ROI), and payback period (PP) are some of the economic metrics that must be evaluated to determine the profitability of pretreatment methods [43].

8.6 SENSITIVITY ANALYSIS

Sensitivity analysis is the process of determining the impact of small deviations from the ideal nominal values of initial conditions, control inputs, and parameters on the process trajectory and the system's results over time, also allowing the investigator to understand the relative importance of these impacts [44]. It can also be used to investigate variations in economic performance and the asymmetric distribution coefficient for different pretreatment techniques, resulting from fluctuations in the cost of raw materials and outputs, which include material exchange within the process plant [45]. Sensitivity analysis examines the changes in feedstock, utility consumption, chemical and utility costs, solvent consumption, volume of acid catalyst used, solvent ratio (percentage of solvent to water), the impact of temperature in the pretreatment, influence of pressure on the pretreatment reactor, pretreatment conversion, and other factors [46]. In this case, each input parameter variable will be varied individually while keeping other crucial parameters constant. The choice of input parameters in the numerical simulation of pretreatment in bioenergy is an important area to investigate. Due to the varying processes and parameters, their selection plays a crucial role in determining the outcome of the process. Therefore, analyzing the sensitivity of the pretreatment model forecasts to variations in input supplies provides a means to improve understanding of the model's response to an unlimited selection of pretreatment parameters [47]. Additionally, sensitivity analysis can reveal that the method may produce the optimal result [48]. In general, sensitivity analysis is required to rank parameters by their importance and fix the nominal values of the smallest influential factors [49].

Sensitivity analysis can be conducted using three major techniques: screening, local sensitivity analysis, and global sensitivity analysis. The screening sensitivity method is employed at the initial stage of analysis to identify and remove parameters that have an insignificant impact on the output, thereby minimizing the model's complexity and computational expenses. This method can be applied to analyze the interaction between input and output and identify the most suitable sensitivity technique [50]. In a local sensitivity analysis technique, only one parameter value is varied, while all other input variables are kept constant [44]. Spearman's rank correlation coefficient and Pearson's correlation coefficient are examples of the most commonly used local sensitivity. It has been observed that this technique is suitable for sensitivity ranking when a single relationship between input and output is considered [51]. Due to the complexity of the pretreatment technique, the time required, and the computational cost of simulating working conditions, the local sensitivity analysis technique can be utilized for optimizing pretreatment technologies [46]. The global sensitivity analysis technique can be used to investigate the impact of multiple input parameters on the model output and examine the effect of the relationship between individual parameters on the model result [52]. This method was reported to be suitable for analyzing uncertainty when the global, quantitative, and model-free approaches were involved. Global, in this case, means that the process considered the whole distribution, quantitative indicates that the sensitivity of the parameter is estimated using a particular sensitivity index, and model-free implies that no assumptions on the model functional interaction with its inputs are important for the sensitivity analysis technique to generate accurate data [53]. Variance-based, moment-independent, and non-parametric methods have been developed for global sensitivity analysis and have been improved recently [50]. These techniques are useful in performing a valid sensitivity analysis when model characteristics include non-linearities and interactions. For clarity of the process, it is essential to consider all possible combinations of input factors as inputs. Analysis of variance (ANOVA) derived from experimental design is an example of global sensitivity analysis. This model provides the analyst with information regarding the influence of input parameters on the process response, including the interactive effects among the factors [54]. Despite the importance of sensitivity analysis in the pretreatment application, there are very limited studies that conduct sensitivity analysis of the pretreatment process. A few available ones are completely focused on local methods, neglecting more exhaustive global methods.

8.7 COMPARATIVE ECONOMIC EVALUATION AND ANALYSIS

The economic evaluation of any developed process can be determined by analyzing the various cost components of the entire process. The primary cost components of biomass pretreatment include biomass, chemicals, equipment, and energy used during the process. The primary aim of the economic evaluation of pretreatment is to assess the production cost and the income generated through product sales or applications [55]. The choice of the optimum pretreatment technique does not depend solely on energy conversion efficiency and biodegradability rate; energetic and economic profits are also important. To achieve economically viable pretreatment, it is crucial to maintain operational profits, minimize energy consumption costs, and optimize maintenance [56,57]. Therefore, selecting a suitable pretreatment technique is crucial to achieving economic viability. The expenses of a pretreatment plant are primarily divided into two categories: capital expenditure (CAPEX) and operational expenditure (OPEX), which must be considered to evaluate the economic feasibility of the process. CAPEX refers to the expenses incurred to set up the facility and other necessary infrastructure. The OPEX includes expenses on raw materials, utilities, labor, maintenance and repairs, administration, and general plant expenses [58]. Technology that requires low OPEX and CAPEX, and can sustain itself to generate profits, is usually regarded as economically viable. The CAPEX and OPEX costs for pretreatment techniques vary across different methods, and the selected method determines the cost feasibility. For instance, pulse-electric field pretreatment requires higher CAPEX due to the need for sophisticated equipment, but the OPEX is lower compared to the acidic pretreatment method, which requires high costs of chemicals, water for washing, and possibly energy for heating [59,60]. Therefore, pretreatment methods that require sophisticated equipment require high CAPEX, whereas those that depend mainly on inputs such as chemicals, water, energy, and microorganisms require higher OPEX [61].

CAPEX is a one-time investment, while OPEX is a recurrent expenditure that depends on the utilization of the plant. CAPEX includes expenses for construction, equipment, engineering, and contingency. Economic evaluations of various biomass pretreatments have been conducted in previous studies. It was noted that the CAPEX required for ammonia fiber explosion, dilute acid, and hot water pretreatment methods were \$25.7 million, \$25 million, and \$4.5 million USD, respectively [62]. It was reported that acidic base pretreatment has higher economic viability when considering the product yield. Combined pretreatment addresses the challenges of individual techniques and can reduce energy costs, but at the expense of higher CAPEX and OPEX [63]. It has been observed that mechanical pretreatment, such as microwave heating, requires a higher CAPEX revenue ratio, regardless of the higher investment [64]. Effective pretreatment and cellulose hydrolysis accounted for around 20% of CAPEX and OPEX in the biorefinery process [65], while another study reported that the CAPEX and OPEX of the pretreatment technique accounted for 40% of the total expenditure [66]. Corn stover pretreated with a two-stage acidic pretreatment was observed to have a pretreatment cost of approximately 13% from CAPEX [67]. pH adjustment is typically required for most chemical pretreatments to minimize the inhibition of methanogenic bacteria, which can result in an additional increase in the pretreatment cost. Some chemical pretreatments are corrosive, which require anti-corrosive equipment, thereby increasing the CAPEX [56]. Lower income and higher CAPEX are the primary challenges that render biorefinery uneconomical in the market.

OPEX comprises energy costs, labor costs, chemicals, indirect costs, and maintenance expenses. The indirect expenses include overhead costs during manufacturing, such as energy for production, taxes, repairs and maintenance, and the cost of equipment. Pretreatment applications to enhance biorefinery process results in an increase in OPEX and maintenance expenses [68]. Pretreatment methods such as homogenization, high temperature, ozonation, and ultrasonication have high OPEX. Likewise, mechanical pretreatment techniques require high energy, which increases the OPEX and makes the method challenging [69]. Homogenization pretreatment is recognized for its ease of operation, leading to higher energy efficiency and lower OPEX. For chemical pretreatment methods, ozone pretreatment requires higher energy and sophisticated equipment, which increases the OPEX of the process, while alkali pretreatment was identified to have lower OPEX [70]. Lime pretreatment is easy and lowers OPEX due to the ease of solvent recovery and making the technique feasible [71]. Biological pretreatment requires low energy, but the cost of the enzyme is relatively higher, which increases the OPEX [63,72]. OPEX for chemical and biological pretreatment is higher due to the cost of chemicals and enzymes; therefore, it is essential to reduce these expenses to encourage the commercialization of biofuel. Pretreatment expenses are not universal and depend on the type of method used and the type of biomass being processed. It is therefore essential to carefully ensure that the selected method is suitable for the respective biomass, thereby ensuring the economic viability of the biorefinery.

8.8 PRETREATMENT OPTIMIZATION

Optimization of pretreatment techniques and biorefinery has become a primary interest of researchers and engineers, as they aim to identify how pretreatment costs affect the biorefinery process and to understand the scenarios under which the process is profitable. These techno-economic optimization analyses can be carried out through qualitative and quantitative techniques [73]. Qualitative approaches are necessary for the project strategy and risk evaluation. Tools like SWOT (strengths, weaknesses, opportunities, and threats) can be used for strategy and capital-spending planning. This tool outlines the qualitative conditions that can help determine the overall merit and limitations of different scenarios. Deterministic and stochastic analysis techniques are examples of quantitative techniques. Sensitivity and scenario are examples of deterministic analysis techniques. The required parameters and cost drivers are considered at a range of values (minimum and maximum conditions). With this tool, the absolute best- and worst-case scenarios are modeled (sensitivity analysis). If there are values with high probability, a parameter can be assigned distinctive values while others remain constant (scenario analysis) [74].

In stochastic techniques, model parameters are assigned a probability distribution, and Monte Carlo simulation is the most common approach in this regard. This technique utilizes uncertain parameters with a probability distribution, where the operator selects values from them and uses these values to calculate the outcome. The major challenge in this method is selecting an appropriate probability range that yields better results and facilitates interpretation [75]. Other researchers and engineers combined techno-economic analysis with LCA to optimize the pretreatment method. A combination of LCA with techno-economic analysis facilitates the evaluation of the environmental impact of the pretreatment method, thereby supporting informed decision-making.

8.9 LCA OF BIOMASS PRETREATMENT

LCA is the cradle-to-grave analysis of a process to investigate the impact of the process, irrespective of the production aim. It involves examining the various environmental influences of the pretreatment method on the ecosystem [76]. LCA involves four stages: goal definition, inventory analysis, impact assessment, and interpretation. Inventory refers to the materials used during the pretreatment process, which includes raw materials, all treatment processes, energy utilization/production, waste released, and other materials that may be required to complete the process. For biomass pretreatment, LCA commonly considers climate change (through CO_2 release), water consumption, energy, aquatic and terrestrial eutrophication, and acidification [77]. LCA is an environmental management tool that can be used to analyze the life cycle of a process according to the ISO 14040 and 14044 frameworks [78]. A detailed review of the LCA of biomass pretreatment can assist in identifying the most suitable pretreatment method for the individual biomass [79].

For instance, LCA was applied to biomass pretreatment processes involving steam, liquid hot water (LHW), organosolv, and dilute acid pretreatment methods. The analysis revealed that LHW, when treated with pressurized deionized water, significantly reduces greenhouse gas (GHG) emissions and releases higher sugars for fermentation into biofuel [76]. LCA of dilute acid, steam explosion, dilute alkali, and hot water pretreatment on rice straw reveals that steam explosion yields the best results, with findings indicating a promising energy yield that serves as a substitute for gasoline [80]. An environmental impact assessment of different chemical pretreatments was conducted, revealing that methanol had the minimal influence on global warming potential, acidification, eutrophication, photochemical oxidation requirements, and marine and human ecotoxicity, while sodium hydroxide pretreatment posed the strongest harmful influence on the ecosystem [81]. Combined pretreatment using mechanical, alkaline peroxide, and enzymatic hydrolysis on sugarcane bagasse was found to reduce energy consumption by 65% with lower waste release compared to traditional chemical pretreatment techniques [82]. LCA with an emphasis on biomass pretreatment is comparatively low compared to other stages of biofuel production. Therefore, there is a need for more studies on the LCA of some established pretreatment methods to obtain holistic information about their efficiency.

8.10 CONCLUSION

Different advancements have been made in the pretreatment of the biomass during the biorefinery process. However, some of the techniques have not produced the expected results due to certain drawbacks. Therefore, process synthesis, simulation, and evaluation are necessary to investigate the commercial-scale biorefinery production process, encompassing various pretreatment methods, and identifying the most suitable and economical techniques. The mass balance and net energy are crucial to the economic viability of pretreatment technology; therefore, it is essential to select a technique with a high mass balance and low energy requirements for economic viability. The cost of equipment and consumables in pretreatment depends on the type of pretreatment, and these factors significantly impact the process economy. Techniques with sophisticated equipment required a higher investment cost, whereas those with solvents, enzymes, and high energy required a higher operating cost. Sensitivity analysis is required for the pretreatment technique to investigate the changes in performance when the input parameters are varied. This will assist in determining the economic viability of the pretreatment application and biorefinery process, as pretreatment is a key component of the process. The economy of the pretreatment method largely depends on selecting the most appropriate technique for each type of biomass.

REFERENCES

[1] R.K. Mishra, V. kumar, P. Kumar, K. Mohanty, Hydrothermal liquefaction of biomass for bio-crude production: A review on feedstocks, chemical compositions, operating parameters, reaction kinetics, techno-economic study, and life cycle assessment, Fuel 316 (2022) 123377. https://doi.org/10.1016/J.FUEL.2022.123377.

[2] A.R.G. da Silva, C.E. Torres Ortega, B.G. Rong, Techno-economic analysis of different pretreatment processes for lignocellulosic-based bioethanol production, Bioresour Technol 218 (2016) 561–570. https://doi.org/10.1016/J.BIORTECH.2016.07.007.

[3] V. Larnaudie, M.D. Ferrari, C. Lareo, Techno-economic analysis of a liquid hot water pretreated switchgrass biorefinery: Effect of solids loading and enzyme dosage on enzymatic hydrolysis, Biomass Bioenergy 130 (2019) 105394. https://doi.org/10.1016/J.BIOMBIOE.2019.105394.

[4] L. Tao, A. Aden, The economics of current and future biofuels, In Vitro Cell Dev Biol—Plant 45 (2009) 199–217. https://doi.org/10.1007/S11627-009-9216-8/FIGURES/10.

[5] A. Azapagic, Sustainability considerations for integrated biorefineries, Trends Biotechnol 32 (2014) 1–4. https://doi.org/10.1016/J.TIBTECH.2013.10.009.
[6] K.O. Olatunji, D.M. Madyira, Comparative analysis of the effects of five pretreatment methods on morphological and methane yield of groundnut shells, Waste Biomass Valor 15 (2024) 469–486. https://doi.org/10.1007/S12649-023-02177-6/FIGURES/5.
[7] M. Rastogi, S. Shrivastava, Recent advances in second generation bioethanol production: An insight to pretreatment, saccharification and fermentation processes, Renew Sustain Energy Rev 80 (2017) 330–340. https://doi.org/10.1016/J.RSER.2017.05.225.
[8] J.B. Holm-Nielsen, E.A. Ehimen, Biorefinery plant design, engineering and process optimisation, Advances in Biorefineries: Biomass and Waste Supply Chain Exploitation (2014) 89–111. https://doi.org/10.1533/9780857097385.1.89.
[9] S.I. Mussatto, G.M. Dragone, Biomass pretreatment, biorefineries, and potential products for a bioeconomy development, Biomass Fractionation Technologies for a Lignocellulosic Feedstock Based Biorefinery (2016) 1–22. https://doi.org/10.1016/B978-0-12-802323-5.00001-3.
[10] R.K. Morgan, Environmental impact assessment: The state of the art, Impact Assess Proj Apprais 30 (2012) 5–14. https://doi.org/10.1080/14615517.2012.661557.
[11] B. Bao, D.K.S. Ng, D.H.S. Tay, A. Jiménez-Gutiérrez, M.M. El-Halwagi, A shortcut method for the preliminary synthesis of process-technology pathways: An optimization approach and application for the conceptual design of integrated biorefineries, Comput Chem Eng 35 (2011) 1374–1383. https://doi.org/10.1016/J.COMPCHEMENG.2011.04.013.
[12] V. Pham, M. El-Halwagi, Process synthesis and optimization of biorefinery configurations, AIChE J 58 (2012) 1212–1221. https://doi.org/10.1002/AIC.12640.
[13] J. Moncada B, V. Aristizábal M, C.A. Cardona A, Design strategies for sustainable biorefineries, Biochem Eng J 116 (2016) 122–134. https://doi.org/10.1016/J.BEJ.2016.06.009.
[14] K.A. Pyrgakis, A.C. Kokossis, A total site synthesis approach for the selection, integration and planning of multiple-feedstock biorefineries, Comput Chem Eng 122 (2019) 326–355. https://doi.org/10.1016/J.COMPCHEMENG.2018.09.003.
[15] K.O. Olatunji, D.M. Madyira, Application of response surface methodology (RSM) for optimizing methane yield of oxidative pretreated Xyris capensis, in: E3S Web of Conferences 433 (2023) 01007. https://doi.org/10.1051/E3SCONF/202343301007.
[16] K.O. Olatunji, D.M. Madyira, N.A. Ahmed, O. Adeleke, O. Ogunkunle, Modeling the biogas and methane yield from anaerobic digestion of Arachis hypogea shells with combined pretreatment techniques using machine learning approaches, Waste Biomass Valor 14 (2023). https://doi.org/10.1007/s12649-022-01935-2.
[17] C.A.C. Alzate, Biorefineries: Design and Analysis (2018). https://doi.org/10.1201/9781315114088.
[18] R. Dickson, J.J. Liu, A strategy for advanced biofuel production and emission utilization from macroalgal biorefinery using superstructure optimization, Energy 221 (2021) 119883. https://doi.org/10.1016/J.ENERGY.2021.119883.
[19] V. Ashokkumar, R. Venkatkarthick, S. Jayashree, S. Chuetor, S. Dharmaraj, G. Kumar, W.H. Chen, C. Ngamcharussrivichai, Recent advances in lignocellulosic biomass for biofuels and value-added bioproducts—A critical review, Bioresour Technol 344 (2022) 126195. https://doi.org/10.1016/J.BIORTECH.2021.126195.
[20] O.S. Mariana, S.T.J. Camilo, C.A.C. Ariel, A comprehensive approach for biorefineries design based on experimental data, conceptual and optimization methodologies: The orange peel waste case, Bioresour Technol 325 (2021) 124682. https://doi.org/10.1016/J.BIORTECH.2021.124682.

[21] J.C. Solarte-Toro, C.A. Cardona Alzate, Sustainability of biorefineries: Challenges and perspectives, Energies 16 (2023) 3786. https://doi.org/10.3390/EN16093786.
[22] I. Davydenko, M. Hopman, R. Fransen, J. Harmsen, Mass-balance method for provision of net zero emission transport services, Sustainability 14 (2022) 6125. https://doi.org/10.3390/SU14106125.
[23] S. Njoku, B. Ahring, H. Uellendahl, Tailoring wet explosion process parameters for the pretreatment of cocksfoot grass for high sugar yields, Appl Biochem Biotechnol 170 (2013). https://doi.org/10.1007/s12010-013-0299-7.
[24] L. Qin, Z.H. Liu, B.Z. Li, B.E. Dale, Y.J. Yuan, Mass balance and transformation of corn stover by pretreatment with different dilute organic acids, Bioresour Technol 112 (2012) 319–326. https://doi.org/10.1016/J.BIORTECH.2012.02.134.
[25] S.O. Dahunsi, S. Oranusi, V.E. Efeovbokhan, Pretreatment optimization, process control, mass and energy balances and economics of anaerobic co-digestion of Arachis hypogaea (Peanut) hull and poultry manure, Bioresour Technol 241 (2017) 454–464. https://doi.org/10.1016/J.BIORTECH.2017.05.152.
[26] I. Davydenko, W.M.M. Hopman, R. van Gijlswijk, A. Rondaij, J. Spreen, Towards Harmonization of Carbon Footprinting Methodologies: A Recipe for Reporting in Compliance with the GLEC Framework, Objectif CO2 and SmartWay for the Accounting Tool BigMile™ (2019). https://doi.org/10.13140/RG.2.2.10883.45601.
[27] B.Z. Li, V. Balan, Y.J. Yuan, B.E. Dale, Process optimization to convert forage and sweet sorghum bagasse to ethanol based on ammonia fiber expansion (AFEX) pretreatment, Bioresour Technol 101 (2010) 1285–1292. https://doi.org/10.1016/J.BIORTECH.2009.09.044.
[28] R.J. Garlock, V. Balan, B.E. Dale, V. Ramesh Pallapolu, Y.Y. Lee, Y. Kim, N.S. Mosier, M.R. Ladisch, M.T. Holtzapple, M. Falls, R. Sierra-Ramirez, J. Shi, M.A. Ebrik, T. Redmond, B. Yang, C.E. Wyman, B.S. Donohoe, T.B. Vinzant, R.T. Elander, B. Hames, S. Thomas, R.E. Warner, Comparative material balances around pretreatment technologies for the conversion of switchgrass to soluble sugars, Bioresour Technol 102 (2011) 11063–11071. https://doi.org/10.1016/J.BIORTECH.2011.04.002.
[29] J.Y. Zhu, W. Zhu, P. Obryan, B.S. Dien, S. Tian, R. Gleisner, X.J. Pan, Ethanol production from SPORL-pretreated lodgepole pine: Preliminary evaluation of mass balance and process energy efficiency, Appl Microbiol Biotechnol 86 (2010) 1355–1365. https://doi.org/10.1007/S00253-009-2408-7/FIGURES/4.
[30] K.O. Olatunji, D.M. Madyira, Effect of acidic pretreatment on the microstructural arrangement and anaerobic digestion of Arachis hypogea shells; and process parameters optimization using response surface methodology, Heliyon 9 (2023) e15145. https://doi.org/10.1016/J.HELIYON.2023.E15145.
[31] S.O. Dahunsi, S. Oranusi, V.E. Efeovbokhan, Optimization of pretreatment, process performance, mass and energy balance in the anaerobic digestion of Arachis hypogaea (Peanut) hull, Energy Convers Manag 139 (2017) 260–275. https://doi.org/10.1016/J.ENCONMAN.2017.02.063.
[32] S. Poulikidou, S. Heyne, M. Grahn, S. Harvey, Lifecycle energy and greenhouse gas emissions analysis of biomass-based 2-ethylhexanol as an alternative transportation fuel, Energy Sci Eng 7 (2019) 851–867. https://doi.org/10.1002/ESE3.315.
[33] W.H. Chen, B.J. Lin, Y.Y. Lin, Y.S. Chu, A.T. Ubando, P.L. Show, H.C. Ong, J.S. Chang, S.H. Ho, A.B. Culaba, A. Pétrissans, M. Pétrissans, Progress in biomass torrefaction: Principles, applications and challenges, Prog Energy Combust Sci 82 (2021) 100887. https://doi.org/10.1016/J.PECS.2020.100887.
[34] H.H. Khoo, C.Y. Koh, M.S. Shaik, P.N. Sharratt, Bioenergy co-products derived from microalgae biomass via thermochemical conversion—Life cycle energy balances

and CO2 emissions, Bioresour Technol 143 (2013) 298–307. https://doi.org/10.1016/J.BIORTECH.2013.06.004.
[35] J.L. Shie, C.Y. Chang, C.S. Chen, D.G. Shaw, Y.H. Chen, W.H. Kuan, H.K. Ma, Energy life cycle assessment of rice straw bio-energy derived from potential gasification technologies, Bioresour Technol 102 (2011) 6735–6741. https://doi.org/10.1016/J.BIORTECH.2011.02.116.
[36] W. Li, M.M. Wright, Negative emission energy production technologies: A techno-economic and life cycle analyses review, Energy Technology 8 (2020) 1900871. https://doi.org/10.1002/ENTE.201900871.
[37] B. Kumar, N. Bhardwaj, K. Agrawal, V. Chaturvedi, P. Verma, Current perspective on pretreatment technologies using lignocellulosic biomass: An emerging biorefinery concept, Fuel Processing Technology 199 (2020) 106244. https://doi.org/10.1016/J.FUPROC.2019.106244.
[38] K.O. Olatunji, N.A. Ahmed, O. Ogunkunle, Optimization of biogas yield from lignocellulosic materials with different pretreatment methods: A review, Biotechnol Biofuels 14 (2021) 1–34. https://doi.org/10.1186/S13068-021-02012-X.
[39] B.O. Abo, M. Gao, Y. Wang, C. Wu, H. Ma, Q. Wang, Lignocellulosic biomass for bioethanol: An overview on pretreatment, hydrolysis and fermentation processes, Rev Environ Health 34 (2019) 57–68. https://doi.org/10.1515/REVEH-2018-0054.
[40] H. Rabemanolontsoa, S. Saka, Various pretreatments of lignocellulosics, Bioresour Technol 199 (2016) 83–91. https://doi.org/10.1016/j.biortech.2015.08.029.
[41] M. del Mar Contreras-Gámez, Á. Galán-Martín, N. Seixas, A.M. da Costa Lopes, A. Silvestre, E. Castro, Deep eutectic solvents for improved biomass pretreatment: Current status and future prospective towards sustainable processes, Bioresour Technol 369 (2023) 128396. https://doi.org/10.1016/J.BIORTECH.2022.128396.
[42] E. Hosseini Koupaie, S. Dahadha, A.A. Bazyar Lakeh, A. Azizi, E. Elbeshbishy, Enzymatic pretreatment of lignocellulosic biomass for enhanced biomethane production-A review, J Environ Manage 233 (2019) 774–784. https://doi.org/10.1016/J.JENVMAN.2018.09.106.
[43] H.I. Ullah, R. Dickson, E. Mancini, A.A. Malanca, M. Pinelo, S.S. Mansouri, An integrated sustainable biorefinery concept towards achieving zero-waste production, J Clean Prod 336 (2022) 130317. https://doi.org/10.1016/J.JCLEPRO.2021.130317.
[44] A. Trucchia, L. Frunzo, Surrogate based Global Sensitivity Analysis of ADM1-based Anaerobic Digestion Model, J Environ Manage 282 (2021) 111456. https://doi.org/10.1016/J.JENVMAN.2020.111456.
[45] R.T.L. Ng, D.K.S. Ng, R.R. Tan, Optimal planning, design and synthesis of symbiotic bioenergy parks, J Clean Prod 87 (2015) 291–302. https://doi.org/10.1016/J.JCLEPRO.2014.10.045.
[46] Z. Barahmand, G. Samarakoon, Sensitivity analysis and anaerobic digestion modeling: A scoping review, Fermentation 8 (2022) 624. https://doi.org/10.3390/FERMENTATION8110624.
[47] A. Boutoute, N. Di Miceli Raimondi, R. Guilet, M. Cabassud, C. Amodeo, H. Benbelkacem, P. Buffiere, R. Teixeira Franco, S. Hattou, Development of a sensitivity analysis method to highlight key parameters of a dry anaerobic digestion reactor model, Biochem Eng J 173 (2021) 108085. https://doi.org/10.1016/J.BEJ.2021.108085.
[48] A. Sipos, P. Agachi, X. Meyer, P. Strehaiano, Batch fermentation process: Modelling and direct sensitivity analysis, Acta Aliment 39 (2010) 222–233. https://doi.org/10.1556/AALIM.39.2010.2.13.
[49] A. Saltelli, K. Aleksankina, W. Becker, P. Fennell, F. Ferretti, N. Holst, S. Li, Q. Wu, Why so many published sensitivity analyses are false: A systematic review of sensitivity

analysis practices, Environ Model Softw 114 (2019) 29–39. https://doi.org/10.1016/J.ENVSOFT.2019.01.012.
[50] D. Li, P. Jiang, C. Hu, T. Yan, Comparison of local and global sensitivity analysis methods and application to thermal hydraulic phenomena, Prog Nucl Energy 158 (2023) 104612. https://doi.org/10.1016/J.PNUCENE.2023.104612.
[51] S.C. Hora, J.C. Helton, A distribution-free test for the relationship between model input and output when using Latin hypercube sampling, Reliab Eng Syst Saf 79 (2003) 333–339. https://doi.org/10.1016/S0951-8320(02)00240-5.
[52] B. Iooss, P. Lemaître, A review on global sensitivity analysis methods, Oper Res/Comput Sci Interfaces Ser 59 (2015) 101–122. https://doi.org/10.1007/978-1-4899-7547-8_5/FIGURES/7.
[53] A. Saltelli, Sensitivity analysis for importance assessment, Risk Anal 22 (2002) 579–590. https://doi.org/10.1111/0272-4332.00040.
[54] K.O. Olatunji, D.M. Madyira, Response surface methodology, multi-objective optimization of biogas, and methane yield of pretreated Arachis hypogea shells, Mater Today Proc 105 (2024) 85–91. https://doi.org/10.1016/J.MATPR.2023.07.351.
[55] B. Kumar, N. Bhardwaj, K. Agrawal, V. Chaturvedi, P. Verma, Current perspective on pretreatment technologies using lignocellulosic biomass: An emerging biorefinery concept, Fuel Process Technol 199 (2020) 106244. https://doi.org/10.1016/J.FUPROC.2019.106244.
[56] A. Gonzalez, A.T.W.M. Hendriks, J.B. van Lier, M. de Kreuk, Pre-treatments to enhance the biodegradability of waste activated sludge: Elucidating the rate limiting step, Biotechnol Adv 36 (2018) 1434–1469. https://doi.org/10.1016/J.BIOTECHADV.2018.06.001.
[57] B. Shrestha, R. Hernandez, D.L.B. Fortela, W. Sharp, A. Chistoserdov, D. Gang, E. Revellame, W. Holmes, M.E. Zappi, A review of pretreatment methods to enhance solids reduction during anaerobic digestion of municipal wastewater sludges and the resulting digester performance: Implications to future urban biorefineries, Appl Sci 10 (2020) 9141. https://doi.org/10.3390/APP10249141.
[58] P. Kaur, H.B. Bohidar, F.M. Pfeffer, R. Williams, R. Agrawal, A comparative assessment of biomass pretreatment methods for the sustainable industrial upscaling of rice straw into cellulose, Cellulose 30 (2023) 4247–4261. https://doi.org/10.1007/S10570-023-05161-4.
[59] D. Carullo, G. Pataro, F. Donsì, G. Ferrari, Pulsed electric fields-assisted extraction of valuable compounds from arthrospira platensis: Effect of pulse polarity and mild heating, Front Bioeng Biotechnol 8 (2020) 551272. https://doi.org/10.3389/FBIOE.2020.551272/BIBTEX.
[60] T. Suopajärvi, P. Ricci, V. Karvonen, G. Ottolina, H. Liimatainen, Acidic and alkaline deep eutectic solvents in delignification and nanofibrillation of corn stalk, wheat straw, and rapeseed stem residues, Ind Crops Prod 145 (2020) 111956. https://doi.org/10.1016/J.INDCROP.2019.111956.
[61] Preethi, M. Gunasekaran, R. Banu. J, Indexing energy and cost of the pretreatment for economically efficient bioenergy generation, Front Energy Res 10 (2023) 1060599. https://doi.org/10.3389/FENRG.2022.1060599/XML.
[62] T. Eggeman, R.T. Elander, Process and economic analysis of pretreatment technologies, Bioresour Technol 96 (2005) 2019–2025. https://doi.org/10.1016/J.BIORTECH.2005.01.017.
[63] M.R. Atelge, A.E. Atabani, J.R. Banu, D. Krisa, M. Kaya, C. Eskicioglu, G. Kumar, C. Lee, Y. Yildiz, S. Unalan, R. Mohanasundaram, F. Duman, A critical review of pretreatment technologies to enhance anaerobic digestion and energy recovery, Fuel 270 (2020) 117494. https://doi.org/10.1016/J.FUEL.2020.117494.

[64] S. Vani, P. Binod, M. Kuttiraja, R. Sindhu, S.V. Sandhya, V.E. Preeti, R.K. Sukumaran, A. Pandey, Energy requirement for alkali assisted microwave and high pressure reactor pretreatments of cotton plant residue and its hydrolysis for fermentable sugar production for biofuel application, Bioresour Technol 112 (2012) 300–307. https://doi.org/10.1016/J.BIORTECH.2012.02.076.
[65] D. Klein-Marcuschamer, B.A. Simmons, H.W. Blanch, Techno-economic analysis of a lignocellulosic ethanol biorefinery with ionic liquid pre-treatment, Biofuels Bioprod Bioref 5 (2011) 562–569. https://doi.org/10.1002/BBB.303.
[66] A.W. Bhutto, K. Qureshi, K. Harijan, R. Abro, T. Abbas, A.A. Bazmi, S. Karim, G. Yu, Insight into progress in pre-treatment of lignocellulosic biomass, Energy 122 (2017) 724–745. https://doi.org/10.1016/J.ENERGY.2017.01.005.
[67] D. Humbird, R. Davis, L. Tao, C. Kinchin, D. Hsu, A. Aden, P. Schoen, J. Lukas, B. Olthof, M. Worley, D. Sexton, D. Dudgeon, Process Design and Economics for Biochemical Conversion of Lignocellulosic Biomass to Ethanol: Dilute-Acid Pretreatment and Enzymatic Hydrolysis of Corn Stover (2011). https://doi.org/10.2172/1013269.
[68] I. Volschan Junior, R. de Almeida, M.C. Cammarota, A review of sludge pretreatment methods and co-digestion to boost biogas production and energy self-sufficiency in wastewater treatment plants, J Water Process Eng 40 (2021) 101857. https://doi.org/10.1016/J.JWPE.2020.101857.
[69] J. Oladejo, K. Shi, X. Luo, G. Yang, T. Wu, A review of sludge-to-energy recovery methods, Energies 12 (2018) 60. https://doi.org/10.3390/EN12010060.
[70] S. Ismail, M.Q. Saharuddin, M.S.M. Zahari, Upgraded seawater-alkaline pre-treatment of lignocellulosic biomass for bio-methane production, Energy Procedia 138 (2017) 372–379. https://doi.org/10.1016/J.EGYPRO.2017.10.390.
[71] A.K. Kumar, S. Sharma, Recent updates on different methods of pretreatment of lignocellulosic feedstocks: A review, Bioresou Bioprocess 4 (2017) 1–19. https://doi.org/10.1186/S40643-017-0137-9.
[72] B. Shrestha, R. Hernandez, D.L.B. Fortela, W. Sharp, A. Chistoserdov, D. Gang, E. Revellame, W. Holmes, M.E. Zappi, A review of pretreatment methods to enhance solids reduction during anaerobic digestion of municipal wastewater sludges and the resulting digester performance: Implications to future urban biorefineries, Appl Sci 10 (2020) 9141. https://doi.org/10.3390/APP10249141.
[73] E. Hytönen, P. Stuart, Techno-economic assessment and risk analysis of biorefinery processes, Compu Aided Chem Eng 29 (2011) 1376–1380. https://doi.org/10.1016/B978-0-444-54298-4.50054-4.
[74] S.I. Mussatto, N. Bikaki, Technoeconomic considerations for biomass fractionation in a biorefinery context, Biomass Fractionation Technologies for a Lignocellulosic Feedstock Based Biorefinery (2016) 587–610. https://doi.org/10.1016/B978-0-12-802323-5.00025-6.
[75] D. Humbird, R. Davis, L. Tao, C. Kinchin, D. Hsu, Process Design and Economics for Biochemical Conversion of Lignocellulosic Biomass to Ethanol, National Renewable Energy Laboratory (2011) 275–3000.
[76] A. Prasad, M. Sotenko, T. Blenkinsopp, S.R. Coles, Life cycle assessment of lignocellulosic biomass pretreatment methods in biofuel production, Int J Life Cycle Assess 21 (2016) 44–50. https://doi.org/10.1007/S11367-015-0985-5/TABLES/8.
[77] R. Shi, R.M. Handler, D.R. Shonnard, Life cycle assessment of novel technologies for algae harvesting and oil extraction in the renewable diesel pathway, Algal Res 37 (2019) 248–259. https://doi.org/10.1016/J.ALGAL.2018.12.005.
[78] M. Finkbeiner, From the 40s to the 70s—The future of LCA in the ISO 14000 family, Int J Life Cycle Assess 18 (2013) 1–4. https://doi.org/10.1007/S11367-012-0492-X/METRICS.

[79] A.T. Ubando, D.R.T. Rivera, W.H. Chen, A.B. Culaba, A comprehensive review of life cycle assessment (LCA) of microalgal and lignocellulosic bioenergy products from thermochemical processes, Bioresour Technol 291 (2019) 121837. https://doi.org/10.1016/J.BIORTECH.2019.121837.
[80] D. Kumar, G.S. Murthy, Life cycle assessment of energy and GHG emissions during ethanol production from grass straws using various pretreatment processes, Int J Life Cycle Assess 17 (2012) 388–401. https://doi.org/10.1007/S11367-011-0376-5/FIGURES/8.
[81] E. Smullen, J. Finnan, D. Dowling, P. Mulcahy, The environmental performance of pretreatment technologies for the bioconversion of lignocellulosic biomass to ethanol, Renew Energy 142 (2019) 527–534. https://doi.org/10.1016/J.RENENE.2019.04.082.
[82] S. Chuetor, V. Champreda, N. Laosiripojana, Evaluation of combined semi-humid chemo-mechanical pretreatment of lignocellulosic biomass in energy efficiency and waste generation, Bioresour Technol 292 (2019) 121966. https://doi.org/10.1016/J.BIORTECH.2019.121966.

PART III

Structural analysis and biomethane modeling approaches

9 Structural analysis of pretreated lignocellulose biomass

9.1 INTRODUCTION

With the recent challenges of energy security and the need to minimize greenhouse gas emissions, the use of biomass has become an important source of sustainable energy [1]. Second-generation biofuels, such as biogas, bioethanol, biodiesel, and biohydrogen, derived from biomass, offer a promising alternative to fossil fuels and present a viable opportunity to reduce the environmental footprint of energy generation [2,3]. Biomass conversion to biofuels generates net-zero greenhouse gas emissions; therefore, biofuels released are carbon-neutral. Biomass is a composition of naturally produced materials like forestry and agricultural residues, livestock waste, energy grass, algae, and aquatic plants. It may also include the organic portion of municipal solid wastes, wastewater sludge, and anthropogenic waste streams [4]. The biomass energy potential is estimated to be approximately 80 times (412 x 10^15 kJ) higher than the energy consumed globally annually [4]. It is regarded as a sustainable and alternative energy source because of its readily available and continuous supply [5]. This material can be converted into biofuels using various technologies, including anaerobic digestion, dark fermentation, gasification, transesterification, fermentation, and pyrolysis [6]. Biomass is a composition of primarily lignin (15–20%), hemicellulose (25–30%), and cellulose (40–50%), and other trace components like nitrogen, pectin, ash, and other inorganic components whose quantity depends on different factors like origin, age, season, etc. Cellulose is a crystalline structure with fibrous characteristics similar to those of lignocellulose, providing a strong base for other components to deposit and interact with it [7]. The arrangement between lignin, hemicellulose, and cellulose led to the complexity and functionality of lignocellulose biomass. However, the hemicellulose gives the matrix its structural stability by withstanding the cellulosic fibers and lignin that are situated between the cellulose and hemicellulose matrix [7]. The intricate morphological structure of biomass components, formed through covalent and non-covalent bonds, presents a recalcitrant challenge for converting biomass into sustainable bioenergy [8].

Hemicellulose and cellulose are carbohydrate components that can be transformed into different forms of bioenergy, while lignin is a renewable source of aromatics in nature that can be utilized as a substrate for generating aromatic compounds [9]. The crystalline arrangement of cellulose, hydrophilic hemicellulose, and cross-bonds of polysaccharide-lignin via ester and ether bonds (covalent bond, phenyl glycosidic, and ferulate acid) of lignocellulose biomass makes it resistant to bioenergy conversion [10]. Therefore, the application of pretreatment is necessary to reduce the

DOI: 10.1201/9781003633815-13

recalcitrant properties of lignocellulose biomass prior to the biorefinery process. It is of paramount importance in improving the efficiency of biomass fractionation and reducing the cost of the biorefinery process. Different types of pretreatment methods, such as biological, thermal, mechanical, chemical, and combined treatments, have been experimented with. The primary focus of pretreatment methods is to increase the surface area, decrease the crystallinity, and decrease the polymerization degree [11]. Pretreatment techniques alter the structural arrangement of biomass, leading to changes in its microstructural and chemical compositions, which vary depending on the specific pretreatment method employed. This chapter provides an overview of some of the important methods available for determining the structural disruption and chemical composition of biomass. The influence of different pretreatment techniques on elemental composition, microstructure, crystallinity index (CrI), and functional groups of different feedstocks is highlighted.

9.2 PROXIMATE AND ULTIMATE ANALYSIS

Proximate and ultimate characteristics of biomass are crucial to the success of the biorefinery process. Proximate characteristics, which include moisture content, total solids (TS), ash, and volatile solids (VS), play a crucial role in the conversion of biomass to biofuel. These can be determined using different standard test procedures, such as the Official Methods of Analysis (AOAC) [12], ASTM E 871–82, and ASTM E872–82 [13,14]. TS is the mass of the solid sample that remained after the removal of the available moisture. It is a composition of both organic and inorganic contents of the biomass. Higher TS beyond this point does not support biofuel biorefinery due to its capacity to increase the viscosity of the process, making it challenging. Likewise, a TS percentage that is too low can dilute the process and lower its efficiency [15]. VS is the percentage of biomass that is organic matter that is available for bioenergy production. This represents primarily the percentage of biomass that can generate biofuel and energy in the case of solid bioenergy. Available studies show that pretreatment techniques have an influence on the TS and VS of biomass. The degree of influence depends on the severity of the pretreatment applied. For example, TS and VS of corn stover were reduced from 91.91 and 89.84 to 89.83 and 88.92%, respectively, when NaOH pretreatment was applied [16]. In a similar study, corn stover pretreated with natural deep eutectic solvents (NADES) improved the TS and VS from 92 and 87 to 93 and 91%, respectively [17]. It can be observed that NaOH pretreatment reduces the TS and VS, while NADES increases the TS and VS when the same feedstocks were considered. This can be attributed to the severity of NaOH in breaking down the structural arrangement of the biomass, resulting in the loss of some corn stover content. However, in the case of NADES pretreatment, it has a mild impact on the biomass, retaining a higher percentage of its original composition. *Opuntia ficus-indica* pretreated with 20% HCl at 80 °C for 1 hr reduced the TS from 6.6% of the untreated to 5.3% after pretreatment, but VS was increased from 78.2% of the untreated to 80.5% in the pretreated sample [18]. DES pretreatment of groundnut shells was observed to reduce the TS for all treatment conditions, whereas the VS was increased for all treatment conditions compared to untreated samples [19]. It was established that the TS and VS of biomass are influenced by pretreatment applications, and the degree of impact depends on the type of pretreatment. The use of strong pretreatment conditions can degrade the biomass to the extent that some TS and VS will be lost. As a result of this and the importance of TS and VS in biorefinery, it is crucial to determine the impact of each treatment on each biomass type to prevent the loss of important constituents.

Proximate characteristics of biomass are crucial information in bioenergy production. This refers to the estimation of the available elements, such as carbon, nitrogen, sulfur, chlorine, hydrogen, and oxygen, present in a sample. These elements play a significant role in biofuel production. For instance, carbon in the feedstock serves as food for microorganisms during anaerobic digestion, while nitrogen is required for their growth. Therefore, a certain amount of these elements is required for efficient biogas production [20]. The ultimate characteristics of biomass are influenced by the type of pretreatment applied, which is determined by the specific pretreatment method used. Pretreatment of *Opuntia ficus-indica* under different treatment conditions was reported to have a significant influence on the elemental composition. Pretreatment using 20% NaOH at 80 °C for 24 hrs produced 34.6, 1.23, and 6.5% C, N, and H, respectively, compared to the untreated sample, which had 32.8, 1.23, and 5.2% C, N, and H, respectively. When the same feedstock was pretreated with 20% HCl for 1 hr at 80 °C, C, N, and H were observed to be 29.4, 0.97, and 7.0%, respectively [18]. It was observed that NaOH increased the C, whereas HCl reduced it; no significant difference in N was observed for NaOH, but it was reduced by HCl. Both pretreatments were observed to increase H, but the increase was higher in HCl than in NaOH. These variations in the influence of pretreatment can be attributed to the strength of each pretreatment in degrading the biomass and losing a percentage of the compositional elements. Loblolly pine was subjected to thermal pretreatment using wet and dry torrefaction at 230 and 250 °C, respectively. Ultimate analysis of the samples was observed to have C—56.05, H—5.94, N—0.09, and O—37.92% for wet torrefaction, and C—50.73, H—6.21, N—0.12, and O—42.94% for dry torrefaction pretreatment, while the untreated sample has C—50.25, H—5.97, N—0, and O—43.34% [21]. It can be observed that wet torrefaction significantly increases the C and N contents, while there is no significant difference in H content, and O content is reduced significantly compared to the raw sample. For dry torrefaction, there is no significant difference in C, while H and N are increased, but O is reduced compared to the untreated sample. It can be established that pretreatment methods and conditions have an influence on the elemental composition of biomass. Therefore, when selecting pretreatment methods and conditions, it is crucial to consider the influence of the selected methods on the elemental composition and to account for the elements required for subsequent bioenergy production.

9.3 MORPHOLOGICAL ANALYSIS

The aim of pretreatment is to disrupt the morphological arrangement of lignocellulose biomass; therefore, it is crucial to investigate the degree of fractionalization that occurs during pretreatment applications. The influence of pretreatment on the morphological arrangement of biomass can be examined using scanning electron microscopy (SEM). This equipment can be used to study the variation between biomass morphology before and after pretreatment. It uses an electron beam to scan the sample and produce a magnified image for assessment [22]. The SEM image reveals damage to lignin and alterations in cell wall arrangement during the examination. Studies have shown that the untreated biomass mostly has smooth and definite arrangements compared to the disrupted and broken structures of pretreated biomass. Observations from SEM images suggest that pretreatment delignification enhances accessibility; however, the degree of delignification varies depending on the specific pretreatment method employed. SEM analysis of $NaCO_3$ pretreated sugarcane bagasse showed a loosened fibrous arrangement with pore appearance, and lignin droplets were noticed, while the untreated substrate had a fine, ordered, and rough appearance [23]. For instance, the effect of different pretreatments on the microstructural arrangement of *Xyris capensis* was investigated under SEM. SEM images of untreated and pretreated samples picked at 200-X magnification are presented in Figure 9.1. In this case, different chemical pretreatment conditions were applied to the same feedstock. Alkali pretreatment conditions, using 2% w/w for 45 min at 90 °C, and acidic pretreatment, using 12% v/v HCl for 15 min at 100 °C, were experimented with. Similarly, thermochemical pretreatment using 5% w/v H_2O_2 for 60 min at 100 °C, while oxidative pretreatment was applied using a Prinha solution of 85% H_2O_2 and 15% H_2SO_4 for 120 min at 90 °C. The impacts of these pretreatments were examined using SEM, and the image is presented in Figure 9.1.

The SEM image of the untreated feedstock reveals a strongly bound surface with a fine-bundle surface structure, featuring various fiber layers that can hinder the degradation of the biomass during the downstream biorefinery process. However, when compared to the alkali, acidic, oxidative, and thermochemical pretreatment samples, it was observed that all the chemical pretreatments applied significantly affect the microstructural arrangement of the biomass, but to different degrees. This supports the report of several studies on the effect of pretreatment techniques on

FIGURE 9.1 SEM images of untreated and chemically pretreated *Xyris capensis* [27].

FIGURE 9.1 ***(Continued)***

the microstructural arrangement of biomass [19,24]. It can be observed that alkali pretreatment resulted in a degraded and chiseled microstructure, which improved the biomass crystallinity and is expected to enhance the availability of the biomass for subsequent bioprocessing (treatment A). The pretreatment applied degraded the amorphous and the carbonyl bonds of the biomass. It was also noticed that the internal tissue of the biomass was highly exposed, which will enhance the downstream process of biorefinery. When acidic pretreatment was applied to the same biomass, different effects were observed compared to alkali pretreatment, as shown in treatment B. It was noticed that the initial fine arrangements were disrupted, loosened, and splinted due to the application of acid. This pretreatment generates a swollen surface with improved fiber disintegration on the biomass surface. Oxidative pretreatment was observed to produce a different impact compared to alkali and acid pretreatment techniques, as shown in Figure 9.1D. Oxidative pretreatment disrupts the cell walls and loosens the particles in the biomass, but to a degree different from other pretreatment methods. The process resulted in the partial degradation of lignin, exposing the internal tissues of the feedstock and making cellulose and hemicellulose available for sugar production. The degree of the biomass defibrillation depends on the percentage of acid (H_2SO_4) in the oxidative solution. The SEM image of the thermochemical pretreatment of *Xyris capensis* is shown in 9.1E, revealing a significant disruption of the feedstock compared to the untreated sample (Figure 9.1A), which differs from other pretreatment techniques. The image depicts a chiseled surface with a clustered, flat-like, small-grained formation, which indicates the degradation of hemicellulose and cellulose to a higher degree, which is expected to improve the downstream biorefinery process. The SEM images from this biomass show that different pretreatment techniques produce varying degrees of impact on a single feedstock. This finding agrees with previous studies that different pretreatment techniques produce varying degrees of impact on a single feedstock [17,25,26].

9.4 CRYSTALLINITY AND CRYSTALLINITY INDEX ANALYSIS

The cellulose content of biomass is composed of recalcitrant polysaccharides that have both amorphous and crystalline regions; however, hemicellulose is mainly amorphous in nature. Due to the crystallinity characteristics of lignocellulose biomass, it exhibits high resistance to conversion during the biorefinery process [28]. The performance of the pretreatment can be determined through the analysis of the crystalline structure of the biomass. The crystalline nature of the biomass can be examined using X-ray diffraction (XRD), which diffracts the sample and produces clear patterns of the biomass, thereby assisting in the study of cellulose microstructure [29]. It can be applied to determine the influence of different pretreatment or treatment conditions on lignocellulose biomass. XRD spectra can exhibit lower and higher peaks corresponding to cellulose crystalline arrangements protected by lignin and hemicellulose contents, respectively, and amorphous cellulose arrangements, respectively [30]. There are various models of XRD equipment available for this analysis, which operate at different temperatures, speeds, and diffraction angles. This equipment offers the opportunity to determine the CrI of the biomass. CrI (%) defines the percentage of crystalline content of the biomass and can be calculated using Equation 9.1 [31].

$$CrI(\%) = \frac{I_{max} - I_{min}}{I_{max}}, \tag{9.1}$$

where CrI (%) = crystallinity index, I_{max} = maximum diffraction at peak position at 2Θ, and I_{min} = the intensity at 2Θ.

The values of I_{max} and I_{min} depend on the type of feedstock and pretreatment applied, but they are typically around 22 °C and 18 °C, respectively [22]. Several studies have used XRD analysis to determine the CrI of different biomasses before and after pretreatment, and some of the results are presented in Table 9.1. The table shows that pretreatments have an influence on the CrI of lignocellulose biomass compared to the untreated biomass. It was observed that the effect varies depending on the type of feedstock, pretreatment method, and treatment conditions. CrI increases in some pretreatments and decreases in others when compared to the untreated samples, confirming that pretreatment does not have a uniform impact on feedstock. The determination of crystallinity in biomass is crucial to biorefinery;

TABLE 9.1
Crystallinity index of some untreated and pretreated biomass

S/N	Biomass	Pretreatment type	Pretreatment conditions	CrI (%)	Ref.
1.	Corn stover	Untreated	Untreated	48.14	[32]
2.	Corn stover	Thermal	Steam explosion at 1.2 MPa for 10 min	57.49	[32]
3.	Corn stover	Thermochemical	KOH at 60 °C for 12 h	58.07	[32]
4.	Corn stover	Combined	Steam explosion + KOH at 60 °C for 12 h	65.52	[32]
5.	Yellow thatching grass	Mechanical	Pulped	70.00	[33]
6.	Yellow thatching grass	Deep eutectic solvent	ChCl: urea	37.00	[33]
7.	Yellow thatching grass	Deep eutectic solvent	Ethylene glycol: citric acid	72.00	[33]
8.	Yellow thatching grass	Deep eutectic solvent	ChCl: glycerol	64.00	[33]
9.	Yellow thatching grass	Untreated	Untreated	73.00	[33]
10.	Groundnut shells	Deep eutectic solvent	ChCl: ethyl glycerol	20.71	[19]
11.	Groundnut shells	Thermal	Autoclave at 120 °C for 30 min	55.26	[25]
12.	Groundnut shells	Alkali	4% w/w NaOH at 90 °C for 10 min	43.66	[34]
13.	Groundnut shells	Acid	0.5% v/v H_2SO_4 at 90 °C for 15 min	59.00	[35]
14.	Groundnut shells	Untreated	Untreated	61.31	[25]
15.	Corn stover	Deep eutectic solvent	1: 2 of NH_4SCN + urea at 80 °C for 1 hr with 1:4 solid–liquid ratio	62.10	[17]
16.	Corn stover	Deep eutectic solvent	1:2 of NH_4SCN + urea at 100 °C for 1 hr with 1:2 solid–liquid ratio	58.40	[17]
17.	*Xyris capensis*	Alkali	4% w/w NaOH at 90 °C for 20 min	31.20	[27]
18.	*Xyris capensis*	Oxidative	85% H_2O_2 + 15% H_2SO_4 for 2 hr at 90 °C	58.00	[36]
19.	*Xyris capensis*	Untreated	Untreated	57.00	[36]
20.	Wheat straw	Ionic liquid	[TEA] [H_2SO_4] for 30 min	66	[37]
21.	Wheat straw	Ionic liquid	[TEA] [H_2SO_4] for 3 hr	71	[37]
22.	Wheat straw	Untreated	Untreated	68	[37]

however, most studies have not considered it, which makes understanding pretreatment technology somewhat challenging, as there is no evidence to show the impact of pretreatment technology on the crystalline structure. Only a few feedstocks have been investigated using XRD after pretreatment, necessitating the need to establish this method for numerous biomass types that have not been considered, for a better understanding of the influence of pretreatment on the feedstock, even before the downstream process of biorefinery.

9.5 FUNCTIONAL GROUP ANALYSIS

Lignocellulose biomass consists of different functional groups, which can be either cellulose, hemicellulose, or lignin, that can be found at different wavelengths. The application of pretreatment techniques has been observed to have a significant influence on these functional groups, which in turn affects subsequent bioenergy production [8]. The influence of pretreatment on these functional groups can be investigated using Fourier transform infrared spectroscopy (FTIR). FTIR can be used to investigate compositional variations in biomass before and after pretreatment applications. It categorizes the chemical links in biomass by generating an infrared absorption spectrum and examining the alterations in biomass functional groups resulting from different pretreatments. Different methods of FTIR spectra analysis, such as multivariate, univariate, and peak fitting, can be used to determine the level of delignification caused by pretreatment [38]. There are various functional groups of lignin, hemicellulose, and cellulose that can be identified at different wavelengths, depending on the type of biomass, pretreatment method, and conditions. Studies have shown that the functional groups of lignocellulose biomass spectra range between 600 and 4000 cm^{-1} with a resolution of 0.5 cm^{-1} [39]. Some of the functional groups include hydroxyl (–OH) straining, –CH vibration of $-CH_2$ and $-CH_3$, symmetric linking in $-CH_2$, symmetric linking C–O–C glycoside, aliphatic C–H distortion stretching or C–H vibration, C–O–C asymmetric vibration, $C-H_2$ wagging, C–O vibration of hemicellulose and cellulose, and C–H asymmetric out of phase ring vibration allocated to cellulose. Other functional groups include asymmetric vibration in –CH3 attached to lignin, carboxylic or acetyl acid (C=O), C=C vibration of aromatic ring, C=C vibration of aromatic ring, C=C aromatic skeletal stretching, and C–O vibration of hemicellulose and cellulose [24,40].

The wavelength, functional group, and assignment for lignocellulose biomass are presented in Table 9.2. The relative variation of the functional group between untreated and pretreated biomass can be determined using Equation 9.2. Positive values of relative variation indicate a decrease, while negative values represent an increase in the functional group [40]. The wavelength for the functional group may differ based on the type of biomass and the pretreatment applied. For instance, the broad bands for the hydroxyl (O–H) group in polysaccharides were discovered at 3330 cm^{-1} for corn stover [17], 3348 cm^{-1} for Sorghum bicolor stalks [40], and 3350 cm^{-1} for *Arachis hypogea* shells [25]. Likewise, the stretching vibration of the C–H link was noticed at 2900, 2904, and 2900 cm^{-1} for corn stover, *Arachis hypogea* shells, and Sorghum bicolor stalks [17,25,40]. There is also the possibility of having varying wavelengths on the same type of feedstock, which could be due to

TABLE 9.2
Wavelengths that correlate to a selected functional group and their response to FTIR

Wavelength (cm^{-1})	Functional Groups	Assignments
3350	O-H stretch	Hydrogen cellulose connections bond
2904	C-H stretch	Methy/methylene cellulose group
1732	Carbonyl bonds	Associated with the removal of lignin side chains
1715	C=C straining vibration of the aromatic ring	Carboxylic acid/ester groups
1635	Aromatic ring stretch	Associated with lignin removal
1606	Aromatic ring stretch	Changes in lignin structure
1515	Generic lignin	
1372	Phenolic O-H stretch	Changes in lignin structure
1320	Syringyl ring stretch	Changes in lignin monomer
1252	C-O absorption	Result of acetyl-lignin groups cleavage
1109	Crystalline cellulose	
1057	C-O-C stretch	Cellulose and hemicellulose
897	Amorphous cellulose	
1109/897	-CH bending vibration in the plane	Crystalline/amorphous cellulose ratio
834	C-H flexion of syringyl	
771	Crystalline cellulose (Iα)	
720	Crystalline cellulose (Iβ)	
771/720	Ratio of crystalline cellulose polymorphs (Iα/Iβ)	

differences in geographical location, age, and time of harvest; however, the variation is expected to be minimal. The wavenumbers of the pretreated and untreated samples differ, depending on the severity of the pretreatment application.

$$Relative\,variation(\%) = \frac{V_u - V_p}{V_u}\ X100, \tag{9.2}$$

where V_u = absorbance of the untreated substrate and V_p = absorbance of the pretreated substrate.

The influence of H_2SO_4 and H_2O_2 on the functional groups of Sorghum bicolor stalks was investigated, revealing that the impacts vary depending on the type of pretreatment. After pretreatment, the O-H vibration was found at 3348 cm^{-1}, and H_2SO_4 increases the O-H vibration (–38.2%), whereas H_2O_2 reduces the O-H vibration (3.2%). Both pretreatment techniques were observed to increase C–H bond strength by 36.3 and 43.9% for H_2SO_4 and H_2O_2, respectively, at 2900 cm^{-1}. Generic lignin at 1516 cm^{-1} was increased by H_2SO_4 but was reduced by H_2O_2. In terms of the crystalline/amorphous cellulose ratio at 1110/897 cm^{-1}, H_2SO_4 increases it, whereas H2O2 pretreatment decreases it compared to the untreated sample [40]. When thermal pretreatment was applied to Arachis hypogea shells at different temperatures and exposure times, the functional group at the respective wavenumber was similar for all the treatment conditions, but with minor differences [25].

9.6 CONCLUSION

This chapter reviews the influence of pretreatment techniques on proximate, ultimate, morphological, crystallinity, and functional groups of biomass. Proximate parameters of TS and VS were noticed to play a significant role in biofuel production, and they are affected by the pretreatment technique. The degree of improvement or reduction depends on the strength of the pretreatment applied, as the degree of degradation of biomass by pretreatment methods results in the loss of TS and VS. Pretreatment methods, such as NADES, were observed to have a mild effect, while strong acids like HCl and H_2SO_4, as well as thermal pretreatment at high temperatures, produce a severe impact on the proximate composition of biomass. Ultimate composition is another important parameter of biomass that influences the efficiency of biorefinery. This is the elemental composition of the biomass, which plays a crucial role in determining the quality and quantity of bioenergy. It is affected by pretreatment techniques, and the severity depends on the type of pretreatment and the treatment conditions. It was also established that the influence is not universal on all biomass, and it needs to be established for each category of biomass. Morphological analysis utilizes SEM to investigate the impact of pretreatment on the microstructural arrangements of biomass, revealing the degree of accessibility of biomass during the biorefinery process. The level of disruption to the microstructural arrangement depends on the type of biomass and the choice of pretreatment. Pretreatment with high penetrating strength was observed to break down the microstructural arrangement more effectively than treatment with low penetrating strength. Crystalline index and functional groups are other important parameters that determine the efficiency of pretreated biomass during biorefinery. These parameters are influenced by the pretreatment methods and vary depending on the specific method and type of biomass. These post-treatment analyses of biomass are crucial to the success of biorefineries, as they provide valuable information that can help optimize the biorefinery process. It was discovered that studies focusing on these parameters are limited; most studies transition directly from pretreatment to the downstream process, which complicates the process. Focusing more attention on post-pretreatment analyses of biomass will deepen the understanding of biomass utilization, as they will provide sufficient information to support downstream bioprocessing and enhance the optimization approach.

REFERENCES

[1] W. Deng, Y. Feng, J. Fu, H. Guo, Y. Guo, B. Han, Z. Jiang, L. Kong, C. Li, H. Liu, P.T.T. Nguyen, P. Ren, F. Wang, S. Wang, Y. Wang, Y. Wang, S.S. Wong, K. Yan, N. Yan, X. Yang, Y. Zhang, Z. Zhang, X. Zeng, H. Zhou, Catalytic conversion of lignocellulosic biomass into chemicals and fuels, Green Energy Environ 8 (2023) 10–114. https://doi.org/10.1016/J.GEE.2022.07.003.

[2] N. Srivastava, R. Singh, P. Singh, I. Ahmad, R.P. Singh, A.K. Rai, M. Asiri, V.K. Gupta, Recent advances on lignocellulosic bioresources and their valorization in biofuels production: Challenges and viability assessment, Environ Technol Innov 29 (2023) 103037. https://doi.org/10.1016/J.ETI.2023.103037.
[3] A.A. Sokan-Adeaga, S.A. Salami, D.O. Bolade, M. Aledeh, M.A. Sokan-Adeaga, O.E. Amubieya, S.A. Kehinde, M. Farzadkia, G.M. Ashraf, E. Hoseinzadeh, Utilization of local corn (Zea Mays) wastes for bioethanol production by separate hydrolysis and fermentation, J Hazard Mater Adv 15 (2024) 100447. https://doi.org/10.1016/J.HAZADV.2024.100447.
[4] A. Tursi, A review on biomass: Importance, chemistry, classification, and conversion, Biofuel Res J 6 (2019) 962–979. https://doi.org/10.18331/BRJ2019.6.2.3.
[5] S. Sharma, P. Nandal, A. Arora, Ethanol production from NaOH pretreated rice straw: A cost effective option to manage rice crop residue, Waste Biomass Valor 10 (2019) 3427–3434. https://doi.org/10.1007/S12649-018-0360-4/TABLES/4.
[6] A.J. Welfle, A. Almena, M.N. Arshad, S.W. Banks, I. Butnar, K.J. Chong, S.J.G. Cooper, H. Daly, S. Garcia Freites, F. Güleç, C. Hardacre, R. Holland, L. Lan, C.S. Lee, P. Robertson, R. Rowe, A. Shepherd, N. Skillen, S. Tedesco, P. Thornley, P. Verdía Barbará, I. Watson, O.S.A. Williams, M. Röder, Sustainability of bioenergy—Mapping the risks & benefits to inform future bioenergy systems, Biomass Bioenergy 177 (2023) 106919. https://doi.org/10.1016/J.BIOMBIOE.2023.106919.
[7] Anu, A. Kumar, D. Singh, V. Kumar, B. Singh, Production of cellulolytic enzymes by Myceliophthora thermophila and their applicability in saccharification of rice straw, Biomass Convers Biorefin 12 (2022) 2649–2662. https://doi.org/10.1007/S13399-020-00783-1/FIGURES/6.
[8] D.M. Madyira, K.O. Olatunji, Investigating the influence of different pretreatment methods on the morphological structure of Arachis hypogea shells, Mater Today Proc 105 (2024) 72–77. https://doi.org/10.1016/J.MATPR.2023.07.289.
[9] J. Zakzeski, P.C.A. Bruijnincx, A.L. Jongerius, B.M. Weckhuysen, The catalytic valorization of lignin for the production of renewable chemicals, Chem Rev 110 (2010) 3552–3599. https://doi.org/10.1021/CR900354U/ASSET/CR900354U.FP.PNG_V03.
[10] Y. Gao, C. Song, M. Usman, Z. Zheng, X. Meng, X. Shen, Y. Cai, X. Wang, Pretreatment of wheat straw using ammonia rich-liquid fraction of digestate: Contribution of biological and non-biological components in methane yield, Ind Crops Prod 220 (2024) 119262. https://doi.org/10.1016/J.INDCROP.2024.119262.
[11] K.O. Olatunji, N.A. Ahmed, O. Ogunkunle, Optimization of biogas yield from lignocellulosic materials with different pretreatment methods: A review, Biotechnol Biofuels 14 (2021) 1–34. https://doi.org/10.1186/S13068-021-02012-X.
[12] Official Methods of Analysis, 21st Edition (2019)—AOAC International (n.d.). https://www.aoac.org/official-methods-of-analysis-21st-edition-2019/ (accessed October 15, 2021).
[13] Test Method for Moisture Analysis of Particulate Wood Fuels (2019). https://doi.org/10.1520/E0871-82R19.
[14] Test Method for Volatile Matter in the Analysis of Particulate Wood Fuels (2024). https://doi.org/10.1520/E0872-24.
[15] L. Yao, Y. Wang, R. Li, L. Fu, Z. Liu, X. Gao, Effects of total solid content on anaerobic fermentation performance and biogas productivity of tail vegetables, Fermentation 10 (2024) 437. https://doi.org/10.3390/FERMENTATION10080437/S1.
[16] E.G. Aklilu, Y.A. Waday, Optimizing the process parameters to maximize biogas yield from anaerobic co-digestion of alkali-treated corn stover and poultry manure using artificial neural network and response surface methodology, Biomass Convers Biorefin 2021 (2021) 1–14. https://doi.org/10.1007/S13399-021-01966-0.

[17] A.D. Olugbemide, A. Oberlintner, U. Novak, B. Likozar, Lignocellulosic corn stover biomass pre-treatment by Deep Eutectic Solvents (DES) for biomethane production process by bioresource anaerobic digestion, Sustainability 13 (2021) 10504. https://doi.org/10.3390/SU131910504.
[18] P.S. Calabrò, E. Catalán, A. Folino, A. Sánchez, D. Komilis, Effect of three pretreatment techniques on the chemical composition and on the methane yields of Opuntia ficus-indica (prickly pear) biomass, Waste Manage Res 36 (2017) 17–29. https://doi.org/10.1177/0734242X17741193.
[19] K.O. Olatunji, D.M. Madyira, Enhancing the biomethane yield of groundnut shells using deep eutectic solvents for sustainable energy production, Front Energy Res 12 (2024) 1346764. https://doi.org/10.3389/FENRG.2024.1346764.
[20] K.O. Olatunji, K.D. Mootswi, O.O. Olatunji, M.I. Zwane, N.J. van Rensburg, D.M. Madyira, Anaerobic co-digestion of food waste and groundnut shells: Synergistic impact assessment and kinetic modeling, Waste Biomass Valor (2025) 1–16. https://doi.org/10.1007/S12649-025-02904-1/TABLES/4.
[21] W. Yan, T.C. Acharjee, C.J. Coronella, V.R. Vá Squez, Thermal pretreatment of lignocellulosic biomass, Environ Prog Sustain Energy 28 (2009) 435–440. https://doi.org/10.1002/ep.10385.
[22] K.O. Olatunji, D.M. Madyira, Comparative analysis of the effects of five pretreatment methods on morphological and methane yield of groundnut shells, Waste Biomass Valor 15 (2024) 469–486. https://doi.org/10.1007/S12649-023-02177-6/FIGURES/5.
[23] Alokika, B. Singh, Enhanced production of bacterial xylanase and its utility in saccharification of sugarcane bagasse, Bioprocess Biosyst Eng 43 (2020) 1081–1091. https://doi.org/10.1007/S00449-020-02306-8/FIGURES/5.
[24] A. Mohammad Rahmani, P. Gahlot, K. Moustakas, A.A. Kazmi, C. Shekhar Prasad Ojha, V.K. Tyagi, Pretreatment methods to enhance solubilization and anaerobic biodegradability of lignocellulosic biomass (wheat straw): Progress and challenges, Fuel 319 (2022) 123726. https://doi.org/10.1016/J.FUEL.2022.123726.
[25] K.O. Olatunji, D.M. Madyira, N.A. Ahmed, O. Ogunkunle, Biomethane production from Arachis hypogea shells: Effect of thermal pretreatment on substrate structure and yield, Biomass Convers Biorefin 14 (2024) 6925–6938. https://doi.org/10.1007/S13399-022-02731-7/METRICS.
[26] X. Li, Y. Shi, W. Kong, J. Wei, W. Song, S. Wang, Improving enzymatic hydrolysis of lignocellulosic biomass by bio-coordinated physicochemical pretreatment—A review, Energy Rep 8 (2022) 696–709. https://doi.org/10.1016/J.EGYR.2021.12.015.
[27] D.M. Madyira, K.O. Olatunji, Alkali pretreatment of lignocellulose feedstock improves morphological structure and biomethane yield, Sustainability 17 (2025) 534. https://doi.org/10.3390/SU17020534.
[28] Z.Z. Rasmeni, D.M. Madyira, A.N. Matheri, O.O. Olatunji, K.O. Olatunji, Mechanical pre-treatment of co-Digested Wastewater Sludge (WWS) and Brewery Spent Yeast (BSY) under anaerobic condition, in: International Conference on Electrical and Computer Engineering Researches, ICECER 2024 (2024). https://doi.org/10.1109/ICECER62944.2024.10920381.
[29] A. Bala, B. Singh, Development of an environmental-benign process for efficient pretreatment and saccharification of Saccharum biomasses for bioethanol production, Renew Energy 130 (2019) 12–24. https://doi.org/10.1016/J.RENENE.2018.06.033.
[30] M. Zheng, X. Li, L. Li, X. Yang, Y. He, Enhancing anaerobic biogasification of corn stover through wet state NaOH pretreatment, Bioresour Technol 100 (2009) 5140–5145. https://doi.org/10.1016/J.BIORTECH.2009.05.045.

[31] R.H. Atalla, D.L. VanderHart, The role of solid state 13C NMR spectroscopy in studies of the nature of native celluloses, Solid State Nucl Magn Reson 15 (1999) 1–19. https://doi.org/10.1016/S0926-2040(99)00042-9.
[32] M.A.H. Siddhu, J. Li, J. Zhang, Y. Huang, W. Wang, C. Chen, G. Liu, Improve the anaerobic biodegradability by copretreatment of thermal alkali and steam explosion of lignocellulosic waste, Biomed Res Int 2016 (2016). https://doi.org/10.1155/2016/2786598.
[33] N.F. Masuku, F. Ayaa, C.M. Onyelucheya, S.A. Iwarere, M.O. Daramola, J.B. Kirabira, Fractionation of yellow thatching grass (Hyparrhenia filipendula) for sugar production using combined alkaline and deep eutectic solvent pretreatment, Waste Biomass Valor 14 (2023) 2609–2620. https://doi.org/10.1007/S12649-022-02013-3/FIGURES/5.
[34] K.O. Olatunji, D.M. Madyira, N.A. Ahmed, O. Ogunkunle, Influence of alkali pretreatment on morphological structure and methane yield of arachis hypogea shells, Biomass Convers Biorefin 14 (2024) 12143–12154. https://doi.org/10.1007/S13399-022-03271-W/METRICS.
[35] K.O. Olatunji, D.M. Madyira, Effect of acidic pretreatment on the microstructural arrangement and anaerobic digestion of Arachis hypogea shells; and process parameters optimization using response surface methodology, Heliyon 9 (2023) e15145. https://doi.org/10.1016/J.HELIYON.2023.E15145.
[36] K.O. Olatunji, D.M. Madyira, Enhancing microstructural and biomethane yield of Xyris capensis using combined oxidative and nanoparticle additive pretreatment methods, Biomass Convers Biorefin 1 (2023) 1–15. https://doi.org/10.1007/S13399-023-04401-8/FIGURES/5.
[37] Z. Ziaei-Rad, J. Fooladi, M. Pazouki, S.N. Gummadi, Lignocellulosic biomass pretreatment using low-cost ionic liquid for bioethanol production: An economically viable method for wheat straw fractionation, Biomass Bioenergy 151 (2021) 106140. https://doi.org/10.1016/J.BIOMBIOE.2021.106140.
[38] M.A.A. Mousa, Y. Wang, S.A. Antora, A.D. Al-qurashi, O.H.M. Ibrahim, H.J. He, S. Liu, M. Kamruzzaman, An overview of recent advances and applications of FT-IR spectroscopy for quality, authenticity, and adulteration detection in edible oils, Crit Rev Food Sci Nutr 62 (2022) 8009–8027. https://doi.org/10.1080/10408398.2021.1922872.
[39] R. Musule, E. Alarcón-Gutiérrez, E.P. Houbron, G.M. Bárcenas-Pazos, M. del Rosario Pineda-López, Z. Domínguez, L.R. Sánchez-Velásquez, Chemical composition of lignocellulosic biomass in the wood of Abies religiosa across an altitudinal gradient, J Wood Sci 62 (2016) 537–547. https://doi.org/10.1007/S10086-016-1585-0/FIGURES/4.
[40] S.O. Dahunsi, A.T. Adesulu-Dahunsi, C.O. Osueke, A.I. Lawal, T.M.A. Olayanju, J.O. Ojediran, J.O. Izebere, Biogas generation from Sorghum bicolor stalk: Effect of pretreatment methods and economic feasibility, Energy Rep 5 (2019) 584–593. https://doi.org/10.1016/j.egyr.2019.04.002.

10 Merits and limitations of lignocellulose pretreatment

10.1 INTRODUCTION

The use of biomass in biorefineries for biofuel production is more challenging than using first-generation feedstocks due to its recalcitrant characteristics. The aim of a biomass biorefinery is to utilize the entire fraction to make the technology more economically feasible. However, it was observed that about 20% of the theoretical yield is released during biorefinery if pretreatment is not applied. The remaining 80% are inaccessible due to the intertwined properties of the biomass [1]. Therefore, pretreatment is required to break down the complex arrangement of biomass through the degradation of intra- and intermolecular linkages between cellulose, hemicellulose, and lignin fractions. Different pretreatment techniques, such as chemical, thermal, biological, and mechanical, have been employed for biomass pretreatment during the biorefinery process [2]. These pretreatment methods behave differently on feedstock; some completely hydrolyze the hemicellulose content, some degrade lignin effectively, whereas others lower the crystallinity and polymerization degree, thereby enhancing the biorefinery process. Biomass pretreatment has enhanced biorefinery technology, increasing its contribution to the production of renewable and sustainable energy. Different pretreatment methods have varying advantages and disadvantages, which need to be addressed to fully realize the potential application of biomass pretreatment. This chapter highlights the merits and limitations of various pretreatment methods, and some propositions are made on how to address the limitations of biomass pretreatment.

10.2 MERITS OF LIGNOCELLULOSE PRETREATMENT

Biomass pretreatment has been identified as a crucial step for the efficient operation of biomass biorefineries. The aim is to disrupt the strong arrangement of lignocellulose, making it accessible during the downstream process. Different types of pretreatment methods have been investigated, including their economic viability, effectiveness, and impact on the final product, as discussed in previous chapters. These pretreatments offer several benefits, with some applicable to all pretreatments and others related to specific treatment methods. The merits of each pretreatment method were discussed in the previous chapters, while the general merits of biomass pretreatment are discussed as follows:

DOI: 10.1201/9781003633815-14

1) Pretreatment encourages the efficient use of biomass for more efficient bioenergy production and sustainable waste management.
2) It breaks the strong biomass arrangement, improves the surface area, and makes cellulose and hemicellulose more available in the downstream process of biorefinery.
3) It can be used to reduce the release of inhibitory compounds that can resist the downstream efficiency of enzymatic hydrolysis and fermentation processes.
4) Efficient pretreatment methods enhance the recovery rate of the important polysaccharides, resulting in the conversion of a larger percentage of biomass into biofuels.
5) Pretreatment can be used for lignin disintegration or solubilization, which can be recovered and used as raw materials for other valuable materials.
6) Pretreatment techniques can make the biorefinery process cost-effective and efficient.
7) Pretreatment is key to making biomass an attractive feedstock for the sustainable generation of bioenergy, substituting fossil fuels and supporting the circular economy approaches.

10.3 LIMITATIONS OF LIGNOCELLULOSE PRETREATMENT

Biomass pretreatment has been observed to improve bioenergy production and present a promising alternative to fossil fuels, while supporting reductions in greenhouse gas emissions, enhancing energy security, and improving waste management. However, the technology is facing different challenges that hinder its scalability and efficiency. The challenges associated with biomass pretreatment technologies are discussed under different categories in this section.

10.3.1 Technological challenges

The complex properties of lignocellulose biomass pose a primary challenge to an effective biorefinery due to the intrinsic resistance resulting from its strong arrangement and the presence of inhibitory components, particularly lignin. Biomass is sourced from various sources, resulting in significant variation in composition. This makes the standardization of pretreatment methods for different biomass categories challenging. Most of the conventional pretreatment methods have been reported to improve bioenergy production from biomass. However, their applications are still limited due to high energy consumption and the generation of inhibitory compounds [3]. Particle size reduction is the most common pretreatment for biomass due to its size after harvesting. However, it does not have the ability to degrade the lignin, which is a major barrier to the accessibility of biomass. This process requires high energy, and with the increase in energy costs, it is not economically sustainable. Therefore, to improve the economy of this process, reducing the energy required and enhancing process efficiency will improve the overall economy of the entire process [4]. The need for special equipment with corrosion resistance, the cost of chemicals, and the need for neutralization are the major challenges associated with chemical pretreatment. The release of inhibitory compounds that negatively impact subsequent downstream processes is another challenge facing chemical pretreatment. Therefore, there is a need to minimize the release of inhibitory compounds to make the technique more efficient [5]. Every component of biomass can be recovered when the organic solvent pretreatment method is applied, which is an advantage in terms of investment cost in the biorefinery approach. Nonetheless, it required special equipment, and a large quantity of downstream waste is generated, which is a major challenge [6]. On the other hand, ionic liquid pretreatment is preferable to organic solvents due to their broad applicability and tolerance to extreme conditions; however, the need for recovery and the associated costs pose a challenge [7].

Compared to other conventional pretreatments, biological pretreatments produce low inhibitory compounds, which have little influence on the downstream process; however, longer exposure time, specific growth conditions, loss of carbohydrates, larger space required, and high cost of enzymes are the major limitations to the technique [5]. Recently discovered non-conventional pretreatment techniques have been observed to overcome some of the challenges associated with conventional pretreatment techniques; however, most of them are still at the laboratory scale. The transition of these new techniques from laboratory to commercial scale is a challenging and time-consuming process. Some of the limitations to the transfer of these techniques to a commercial scale are the following:

1. The availability of special laboratory equipment for some of these techniques, such as electron beam irradiation, pulse-electric field, plasma pretreatment, etc., is limited and does not allow many researchers to engage in studies that can support their industrial application on time.
2. Compared to the conventional pretreatment methods, the available research outputs are small. Few research groups are investigating specific methods with a small number of biomasses and very low optimization of process parameters.
3. Most of the researchers who are working on non-conventional pretreatment hardly consider environmental sustainability and techno-economic analysis of the technique.
4. The low level of multidisciplinary research group collaborations limits the exchange of ideas and the potential of purpose-fit equipment design and process optimization.

10.3.2 Environmental Challenges

The focus of biorefinery is to produce bioenergy that can help reduce greenhouse gas emissions; however, an inefficient pretreatment process can create additional environmental challenges. For instance, when toxic chemicals are used during pretreatment, they release harmful waste that is detrimental to the environment. Therefore, eco-benign and sustainable pretreatment requires green solvents that align with the circular economy approach [8]. The release of carbon dioxide during pretreatment is another major environmental challenge that needs to be addressed.

Some pretreatment techniques require higher energy, and these energy sources may release CO_2 [9]. Thus, the use of technologies with a minimal carbon footprint is crucial to reducing the environmental impact of pretreatment [10]. Some chemical pretreatments require washing after treatment, and management of this wastewater is a major challenge that can negatively impact the environment when not properly managed [11,12].

10.3.3 Economic challenges

The cost of biofuel production is observed to be high, which makes it difficult to compete with fossil fuels; pretreatment is a major contributor to this high cost. Some pretreatment methods require chemicals and/or high temperatures and pressures, which increase the operational and total production costs. Therefore, it is a challenge to select an economically viable pretreatment method during bioprocessing. Another challenge is the requirement of advanced equipment and technology, which leads to higher investment costs. Recovery of this huge investment cost makes the process economically challenging, considering the lifespan of some of this equipment. The use of biological pretreatment has been observed to improve bioenergy production; however, the high cost of enzymes and microorganisms makes the method challenging. Despite successes recorded in enzyme engineering and biotechnology, the cost of biocatalysts remains high.

10.4 FUTURE PROSPECTS OF LIGNOCELLULOSE PRETREATMENT

To overcome energy and environmental challenges posed by total reliance on fossil fuels and waste management, developing a sustainable biofuel generation technology through second-generation biofuels from biomass is crucial [13]. Despite different technological, environmental, and economic challenges associated with biomass pretreatment that are limiting the adoption of biofuel production from lignocellulose biomass, it still has a significant potential as an alternative to fossil fuels [14]. The success of this technology relies heavily on continuous research and development that will produce more cost-effective and efficient bioenergy from biomass. Bioenergy generation from biomass using innovative pretreatment methods will contribute significantly to the reduction of greenhouse gas release, enhance sustainable waste management, and achieve a circular economy [15].

Findings from this study indicate that future work should focus on a detailed investigation of understudied eco-benign pretreatment methods. Further studies should focus on optimizing the total pretreatment parameters using the most readily available biomass. Studies should focus more on developing new, novel, and low-energy-intensive pretreatment techniques. Further development of microorganisms that can be used at a commercial scale will make the pretreatment of biomass more acceptable [16]. The application of pressure technologies in biomass pretreatment can significantly improve the biorefinery efficiency and encourage the adoption of pretreatment on a commercial scale [17]. Therefore, there is a need to further investigate this technique. The use of sustainability tools, such as techno-economic analysis, life cycle assessment, energy analysis, and exergy analysis, should be more of an interest when investigating the potential of new pretreatment techniques [18]. These tools will help provide clear information about the feasibility and sustainability of new pretreatment techniques being investigated [19]. Investigations on multi-biorefinery need to be intensified because it will improve the overall economy of pretreatment and make it more eco-benign [20]. Multidisciplinary research on pretreatment techniques, involving engineers, bioprocess technologists, biochemists, biologists, and economists, should be encouraged to facilitate holistic research that will make the biomass pretreatment process more economical and environmentally friendly. Further studies are also needed on some of the pretreatment methods that have limited information, and methods that are not sustainable need to be clearly highlighted to save time and cost.

10.5 CONCLUSION

Biomass pretreatment is an indispensable stage in biorefineries to improve process efficiency and encourage the use of biofuel as an alternative energy source. Different pretreatment techniques have been experimented with and reported to improve the efficiency of biomass biorefineries. It was noted that various merits of biomass pretreatment were identified in this chapter; however, several limitations to its application were also highlighted. To make biomass pretreatment and biorefinery technology, further studies are required on existing pretreatment methods and the search for new, novel methods that are more efficient and economical.

REFERENCES

[1] N. Akhtar, K. Gupta, D. Goyal, A. Goyal, Recent advances in pretreatment technologies for efficient hydrolysis of lignocellulosic biomass, Environ Prog Sustain Energy 35 (2016) 489–511. https://doi.org/10.1002/EP.12257.
[2] K.O. Olatunji, N.A. Ahmed, O. Ogunkunle, Optimization of biogas yield from lignocellulosic materials with different pretreatment methods: A review, Biotechnol Biofuels 14 (2021) 1–34. https://doi.org/10.1186/S13068-021-02012-X.
[3] A. Woźniak, K. Kuligowski, L. Świerczek, A. Cenian, Review of lignocellulosic biomass pretreatment using physical, thermal and chemical methods for higher yields in bioethanol production, Sustainability 17 (2025) 287. https://doi.org/10.3390/SU17010287.
[4] S.O. Jekayinfa, A.O. Adebayo, O.O. Oniya, K.O. Olatunji, Comparative analysis of biogas and methane yields from different sizes of groundnut shell in a batch reactor at mesophilic temperature, J Energy Res Rev 5 (2020) 34–44. https://doi.org/10.9734/jenrr/2020/v5i130140.
[5] S.R. Paudel, S.P. Banjara, O.K. Choi, K.Y. Park, Y.M. Kim, J.W. Lee, Pretreatment of agricultural biomass for anaerobic digestion: Current state and challenges, Bioresour Technol 245 (2017) 1194–1205. https://doi.org/10.1016/j.biortech.2017.08.182.
[6] L. Capecchi, M. Galbe, L. Barbanti, O. Wallberg, Combined ethanol and methane production using steam pretreated sugarcane bagasse, Ind Crops Prod 74 (2015) 255–262. https://doi.org/10.1016/J.INDCROP.2015.05.016.
[7] Z. Usmani, M. Sharma, P. Gupta, Y. Karpichev, N. Gathergood, R. Bhat, V.K. Gupta, Ionic liquid based pretreatment of lignocellulosic biomass for enhanced bioconversion, Bioresour Technol 304 (2020) 123003. https://doi.org/10.1016/J.BIORTECH.2020.123003.

[8] M. Mujtaba, L. Fernandes Fraceto, M. Fazeli, S. Mukherjee, S.M. Savassa, G. Araujo de Medeiros, A. do Espírito Santo Pereira, S.D. Mancini, J. Lipponen, F. Vilaplana, Lignocellulosic biomass from agricultural waste to the circular economy: A review with focus on biofuels, biocomposites and bioplastics, J Clean Prod 402 (2023) 136815. https://doi.org/10.1016/J.JCLEPRO.2023.136815.
[9] J. Fan, Y. Lu, N. An, W. Zhu, M. Li, M. Gao, X. Wang, C. Wu, Y. Wang, Pretreatment technologies for lignocellulosic biomass: Research progress, mechanisms, and prospects, Bioresources 20 (2025) 4897–4924. https://ojs.bioresources.com/index.php/BRJ/article/view/24189 (accessed August 25, 2025).
[10] M. Kelbert, T.O. Machado, P.H.H. Araújo, C. Sayer, D. de Oliveira, P. Maziero, K.E. Simons, B.A.M. Carciofi, Perspectives on biotechnological production of butyric acid from lignocellulosic biomass, Renew Sustain Energy Rev 202 (2024) 114717. https://doi.org/10.1016/J.RSER.2024.114717.
[11] S. Meenakshisundaram, A. Fayeulle, E. Léonard, C. Ceballos, X. Liu, A. Pauss, Combined biological and chemical/physicochemical pretreatment methods of lignocellulosic biomass for bioethanol and biomethane energy production—A review, Appl Microbiol 2 (2022) 716–734. https://doi.org/10.3390/APPLMICROBIOL2040055.
[12] O. Sarkar, U. Rova, P. Christakopoulos, L. Matsakas, Organosolv pretreated birch sawdust for the production of green hydrogen and renewable chemicals in an integrated biorefinery approach, Bioresour Technol 344 (2022) 126164. https://doi.org/10.1016/J.BIORTECH.2021.126164.
[13] A.T. Ubando, C.B. Felix, W.H. Chen, Biorefineries in circular bioeconomy: A comprehensive review, Bioresour Technol 299 (2020) 122585. https://doi.org/10.1016/J.BIORTECH.2019.122585.
[14] M. Guo, W. Song, J. Buhain, Bioenergy and biofuels: History, status, and perspective, Renew Sustain Energy Rev 42 (2015) 712–725. https://doi.org/10.1016/J.RSER.2014.10.013.
[15] A.T. Ubando, W.H. Chen, D.A. Hurt, A. Conversion, S. Rajendran, S.L. Lin, Biohydrogen in a circular bioeconomy: A critical review, Bioresour Technol 366 (2022) 128168. https://doi.org/10.1016/J.BIORTECH.2022.128168.
[16] H. Wang, X. Cheng, Z. Li, L. Jing, J. Hu, Photothermal catalytic enhancement of lignocellulosic biomass conversion: A more efficient way to produce high-value products and fuels, Chem Eng J 496 (2024) 153772. https://doi.org/10.1016/J.CEJ.2024.153772.
[17] A. Ramos, E. Monteiro, A. Rouboa, Biomass pre-treatment techniques for the production of biofuels using thermal conversion methods—A review, Energy Convers Manag 270 (2022) 116271. https://doi.org/10.1016/J.ENCONMAN.2022.116271.
[18] M.A. Rosen, Environmental sustainability tools in the biofuel industry, Biofuel Res J 5 (2018) 751–752. https://doi.org/10.18331/BRJ2018.5.1.2.
[19] S. Soltanian, M. Aghbashlo, F. Almasi, H. Hosseinzadeh-Bandbafha, A.S. Nizami, Y.S. Ok, S.S. Lam, M. Tabatabaei, A critical review of the effects of pretreatment methods on the exergetic aspects of lignocellulosic biofuels, Energy Convers Manag 212 (2020). https://doi.org/10.1016/J.ENCONMAN.2020.112792.
[20] M. Aghbashlo, Z. Khounani, H. Hosseinzadeh-Bandbafha, V.K. Gupta, H. Amiri, S.S. Lam, T. Morosuk, M. Tabatabaei, Exergoenvironmental analysis of bioenergy systems: A comprehensive review, Renew Sustain Energy Rev 149 (2021) 111399. https://doi.org/10.1016/J.RSER.2021.111399.

11 Modeling of bioenergy yield of pretreated lignocellulose feedstocks

11.1 INTRODUCTION

Energy insecurity and environmental challenges are among the major global challenges, particularly for developing countries. The energy-environmental nexus approaches are complex, involving human health, ecosystems, and economic development of countries and corporations [1]. These nexus challenges can be addressed using bioenergy production from biomass. The attention of researchers and policymakers has shifted to generating bioenergy from various biomass sources to replace fossil fuels. Bioenergy production from biomass is challenging because of multiple factors and uncertainties [2]. With these factors and uncertainties, along with the complex information, it is challenging to handle the bioenergy production information that supports informed decision-making. Therefore, the acceptance of bioenergy has recently faced uncertainty in decision-making, necessitating an effective technique for informed decision-making. Finding techniques to manipulate data and address relationship problems is discovered through the application of modeling methods (stochastic or deterministic) that allow "what if" questions to be answered. Bioenergy modeling has become a major research interest in bioenergy studies recently [3]. It focuses on different models that aim at analyzing the capacity of bioenergy, one or more important stages, and its addition to the bioenergy system [4]. Bioenergy production efficiency is determined by various input parameters, including pretreatment, temperature, pH, and concentration; therefore, it is essential to determine the optimal values for these parameters for bioenergy generation development and commercial applications [5]. Bioenergy optimization requires varying one or more input parameters while keeping at least one of the input parameters constant. This method is challenging and may not reveal the total impact of the input parameters, as it overlooks the combined interaction between them [6].

Mathematical, statistical, and artificial intelligence (AI) models can reveal important data for the detailed analysis and prediction of the biorefinery process. They are needed to optimize crucial input variables to improve the process yield [7]. It aims to reveal the response of feedstock and input parameters in bioenergy release, providing an in-depth understanding of individual parameters and the interactive influence of various process parameters on expected yield [8]. The selection of bioenergy technology and modeling process will influence the forms and levels of bioenergy that can be produced, as well as the determination of the environmental and economic performance of the bioenergy released. Different models have been developed and applied in biorefineries over the past few years, enabling them to

DOI: 10.1201/9781003633815-15

support decision-making processes [9]. Therefore, this chapter reviews the application of various models, including statistical, AI, and kinetic models, in biorefinery for pretreatment applications as process parameters, providing concise, detailed, and thorough information that can guide the choice of models for bioenergy optimization, especially when pretreatment is involved. This chapter will enhance the upscaling of the biorefinery process by providing baseline information that will assist in selecting the optimum parameters.

11.2 RESPONSE SURFACE METHODOLOGY (RSM)

Optimization of bioenergy requires an accurate choice of input parameters to achieve the optimization goal. Having a suitable and reliable bioenergy optimization model that can forecast bioenergy generation accurately, instead of the experimental optimization process, which is highly challenging due to the non-linearities of the process [10]. Mathematical and statistical models can reveal important insights for optimizing, designing, and analyzing bioenergy production. RSM is a statistical and mathematical method that can be used to optimize and forecast parameters that can be replicated on an industrial scale and assist in calculating and determining the depth dynamics of the biorefinery process using simple and generalizable data optimization [11]. It has been noticed for its capacity to evolve, enhance, and optimize the complex biorefinery process [12]. This model has been found to be suitable for optimizing processes that require experimental design, such as chemical reactions, agricultural processes, and industrial processes [13]. Additionally, it can generate a mathematical equation that can be used to analyze the influence of independent variables. Optimal process conditions could be either the lowest or the highest for a specified function in process input variables.

The RSM model can define the interaction between one or more ascertained responses and the crucial controllable biorefinery input variables. If all the independent input variables are measurable and replicated with a small error margin, the model can be presented as illustrated in Equation 11.1 [14].

$$m = f\left(a_1, a_{2,} \ldots\ldots\ldots\ldots\ldots\ldots a_z\right), \tag{11.1}$$

where z is the number of dependable inputs.

To optimize "m," there is a need to determine a suitable estimation for the functional relationships between the independent variables and the response output. This can be determined using the second-polynomial equation, as represented in Equation 11.2 [15].

$$m = r_o + \sum r_i a_i + \sum r_{ij} a_i a_j + \sum r_{ii} a_{ii}^2 + \varepsilon. \tag{11.2}$$

Central composite design (CCD) and Box–Behnken design (BBD) are the two most common experimental designs for the design of the second-order polynomial of the RSM. Most biorefinery studies employed CCD, and the model can be analyzed with Equations 11.3–11.9 [16].

$$Sum\,of\,Squares - Total = SS_T = \sum_{i=1}^{n}(a_i - \bar{a})^2, \tag{11.3}$$

$$Sum\,of\,Squares - Model = SS_M = \sum_{i=1}^{n}(\hat{a}_i - \bar{a})^2, \tag{11.4}$$

$$Sum\,of\,Squares - Residuals = SS_R = \sum_{i=1}^{n}(a_i - \hat{a}_i)^2, \tag{11.5}$$

$$Sum\,of\,Squares - Pure\,Error = SS_{PE} = \sum_{i=1}^{n_o}(a_i - \hat{a}_i)^2, \tag{11.6}$$

$$SS_{ri} = r_i \sum_{i=1}^{n} a_i m_i, \tag{11.7}$$

$$SS_{r_{ij}} = r_{ij} \sum a_i a y_j m_i, \tag{11.8}$$

$$SS_{r_{ii}} = r_o \sum_{i=1}^{n} m_i + r_{ii} \sum_{i=1}^{n} a_i^2 m_i - \left(\sum m_i\right)^2 / n. \tag{11.9}$$

The suitability of the RSM model developed can be evaluated using analysis of variance (ANOVA), and the statistical adequacy of the model and regression expression terms can be checked with the sequential Fisher test (F-test) and the lack-of-fit test. Coefficients of correlation (R^2), adjusted R^2, predicted R^2, and adequate precision ratio are other performance metrics used to test the accuracy of the RSM model. The probability of the model can be examined using ANOVA, also referred to as the p-value, for each input parameter. In a condition where the model's probability and that of individual terms are not beyond the 95% significance level ($\alpha = 0.05$), the model is adequate at a 1-α confidence level. When the lack of fit exceeds the significance level, the lack of fit can be reported as insignificant. The ANOVA defines the strength of the model and can be estimated with Equations 11.10–11.14

$$Correlation\,Coefficient\left(R^2\right) = 1 - \left[\frac{SS_R}{SS_R + SS_M}\right], \tag{11.10}$$

$$AdjR^2 = 1 - \left[\left(\frac{SR_R}{df_R}\right) X \left(\frac{SS_R + SS_M}{df_R + df_M}\right)^{-1}\right], \tag{11.11}$$

$$Pred\,R^2\,1\ - \left[\frac{PRESS}{SS_R + SS_M}\right], \tag{11.12}$$

$$PRESS = \sum_{i=1}^{n}\left(y_i - \hat{y}_{i,-i}\right)^2, \tag{11.13}$$

$$Adeq.\,precision = \left[\frac{Max(\hat{y}) - Min(\hat{y})}{\sqrt{\frac{p\,X\,MS_R}{n}}} \right]. \tag{11.14}$$

The interactive effect of input parameters can be studied through a three-dimensional (3D) surface and contour plot generated from the RSM, as illustrated in Figure 11.1. In this case, one or more parameters will be held at an optimal value, while others are varied across the data ranges until optimal conditions are found. Findings from this figure will interpret the relationship between different variables and their influence on the bioenergy yield and can be validated by the Fisher test value (F-value) obtained from the ANOVA table. Other features that can be used to test the accuracy and optimal conditions of the biorefinery process include perturbation analysis plots,

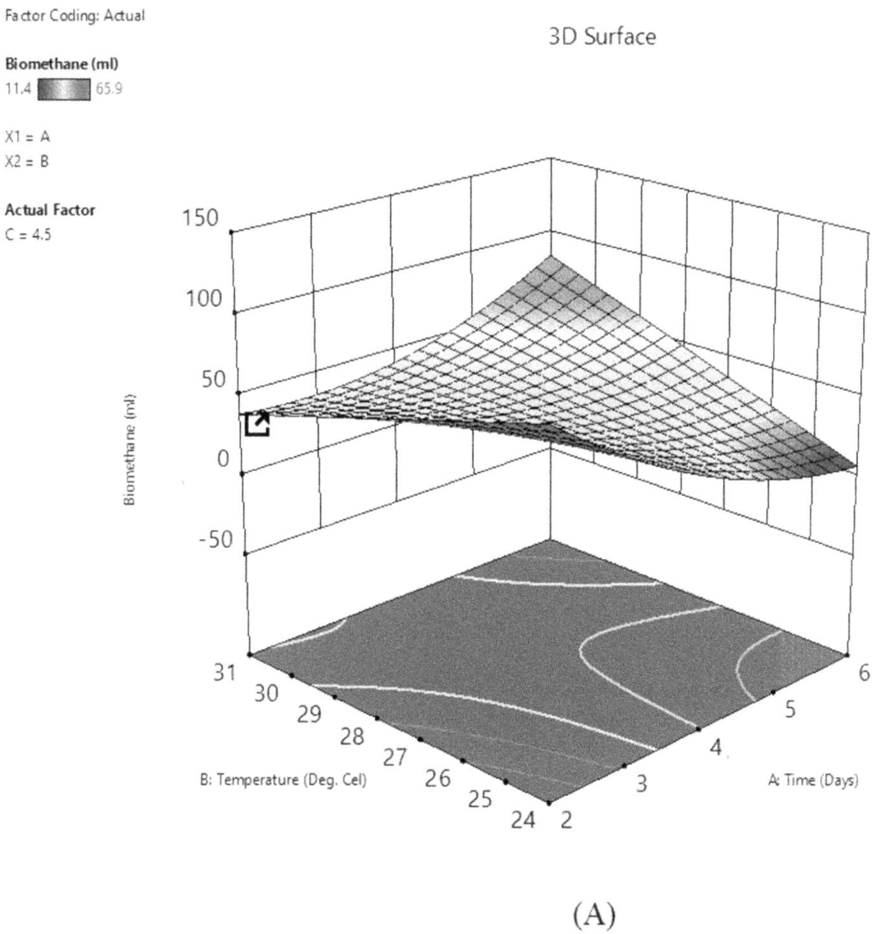

FIGURE 11.1 RSM surface plots: (A) 3D surface plot and (B) contour plot [20].

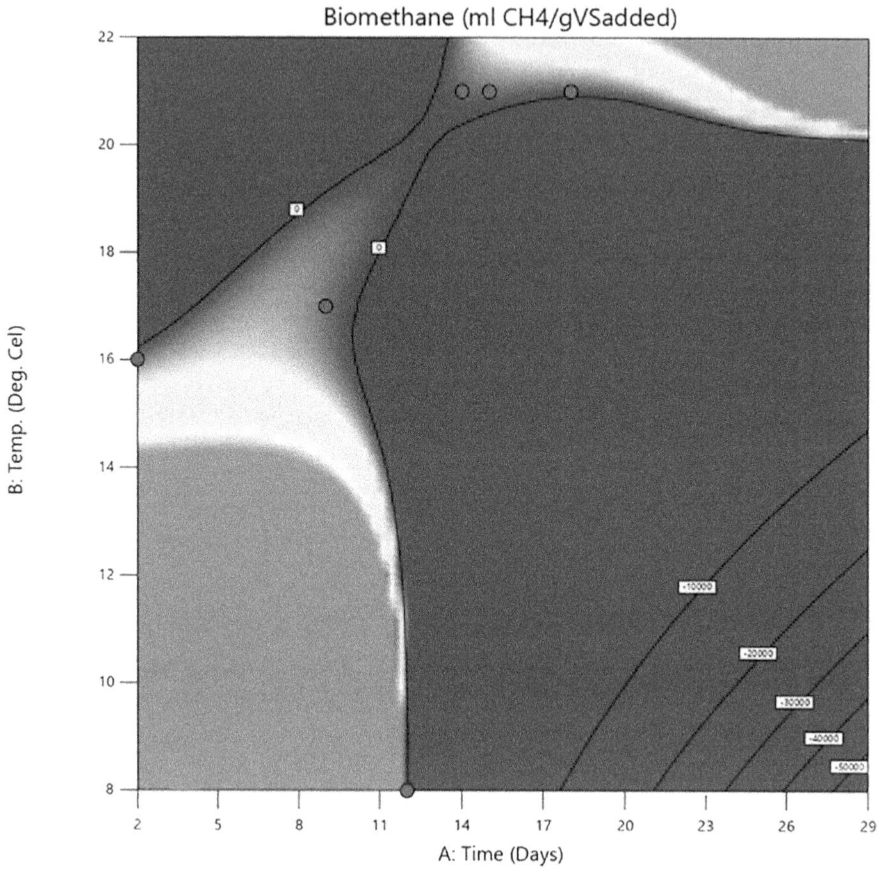

(B)

FIGURE 11.1 (Continued)

desirability analysis plots, Box-Cox plots, Cook's distance plots, cube plots, interaction plots, and ramps plots, among others. Different studies have investigated the use of RSM in optimization and prediction of bioenergy, and it was reported not to have a unified accuracy [17–19]. However, it was observed that very limited studies consider pretreatment as their input parameter. Considering the role pretreatment plays in the biorefinery of biomass and the optimization achieved using the RSM model, there is an urgent need for further studies that focus on the use of RSM for bioenergy optimization and forecasting when the input variable of pretreatment is involved. This will reduce the time and cost of the experimental process that is required for bioenergy commercialization and encourage the scaling up of the biorefinery process.

11.3 ARTIFICIAL INTELLIGENCE MODELS

The biorefinery process from biomass has taken a new dimension, shifting from the use of traditional statistical techniques to the application of intelligent methods in the field of artificial intelligence (AI). AI models have been identified as promising models for the biorefinery process [21]. This has enhanced data-driven decision-making in biorefinery at the strategic and planning phases. This has been observed to enhance bioenergy generation efficiency and stable operation, and it also improves the understanding of the input variables for optimal yield and production rate [22]. The AI models have enhanced the reliability of modeling complex and non-linear connections between inputs and responses, thereby revealing hidden patterns within the datasets [23]. Artificial neural networks (ANNs), particle swarm optimization (PSO), adaptive neuro-fuzzy inference systems (ANFISs), genetic algorithms (GAs), fuzzy logic, and ant algorithms are some of the AI models that have been applied in the biorefinery process [24,25]. These models are categorized into four learning methods: unsupervised, supervised, reinforcement, and ensemble learning [26]. ANN and ANFIS models are examples of supervised AI models and will be overviewed in this section.

11.3.1 Artificial Neural Network

An artificial neural network (ANN) model is one of the most popular AI models identified as a crucial tool for optimizing and predicting bioenergy production [27]. It is an information-processing pattern simulated using the biological nervous system, following the process of how the human brain processes information. It can be defined as a mathematical understanding of neuroscience that mimics the human brain by arithmetically modeling the network arrangement of the interrelated structure of nerve cells [28]. It is fast, efficient, precise, and capable of executing complex tasks without having prior knowledge of the task. Its strength as a model stems from the exceptional information-processing properties, primarily attributed to non-linearity, fault and noise acceptance, generalization strengths, and high parallelism. Compared to conventional models, it features an adaptive, model-free, parallel-processing, and robust design with strong clarity, as well as tolerance to failure and error [21]. It has the capacity to learn and process non-linear data and fuzzy information, and it also identifies general hidden patterns effectively [29]. It is a data-driven model that has an input layer, one or more hidden layers, and an output layer, as illustrated in Figure 11.2 [16]. The hidden layer can be one or more layers that assist the network in defining the complicated interactions between input variables and the response. The output is estimated by adding the non-linear functions of the input variables.

During ANN model development, different observed input parameters, categorized as input parameters, are fed into the model to produce a mathematical relationship between the expected response, called the output, without requiring a priorly identified relationship. The input and output variables (independent and dependent) are interrelated within the model through the nodes dominated neurons (Neus). Neurons receive, store, and process one or more input signals at different strengths based on the influence of input data, and they support the output data with the help of a transfer function [29]. The data can be divided into training, testing, and validation sets, where the training set comprises most of the data. The model can be trained with different learning algorithms, such as scaled conjugate gradient (trainscg), Levenberg-Marquardt (trainlm), gradient descent with adaptive algorithm (traingda), etc., while losgig, tansig, purelin, and softmax can be the activation functions that can be investigated at hidden and output layers [30]. The most suitable and accurate

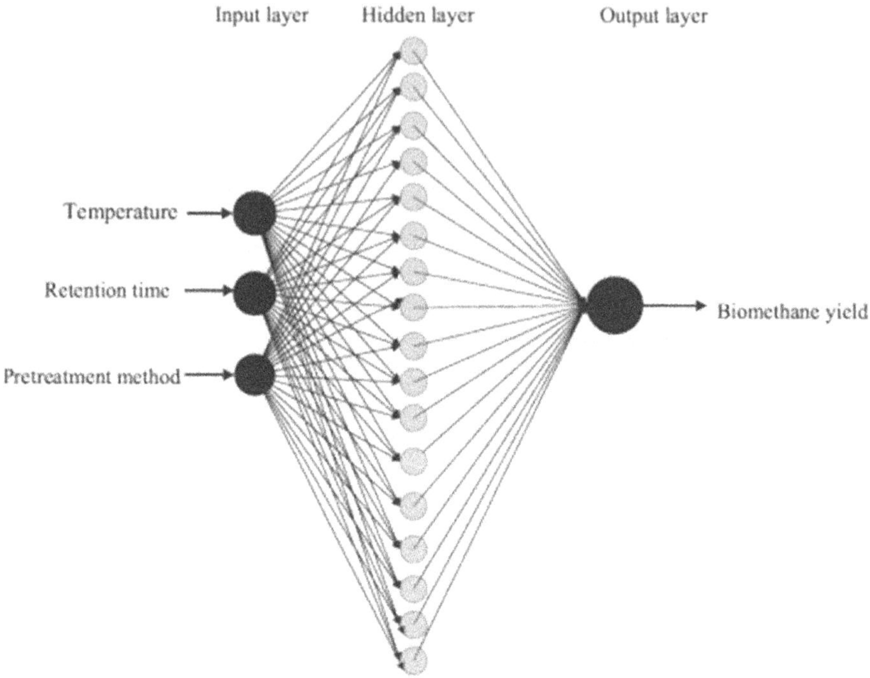

FIGURE 11.2 Architecture of the ANN model [16].

model, along with its corresponding parameters, can be chosen for each model based on the model with the lowest error criteria when examining the performance metrics. The strength of the ANN model can be enhanced by utilizing an error-correction learning algorithm, such as a feedforward backpropagation neural network (FF-BPNN). The neural network can be illustrated mathematically as shown in Equation 11.15, where biases and weights assigned to each layer are adjusted [31]. ANN model is known for their ability to predict, estimate, or make decisions in several fields of engineering and science, including bioenergy production. Several studies have investigated the use of ANN in biorefinery [32–35]. However, there are limited studies that considered pretreatment methods as input variables. Findings from studies that consider pretreatment as input indicate that the performance of the ANN model depends on the pretreatment methods and the expected output [16,30]. Therefore, further studies are required on several pretreatment methods and bioenergy types that are currently lacking in the literature to establish the potential of the ANN model for optimizing and predicting the upscaling of biorefineries.

$$b = f\left(\sum_{i=0}^{m} a_i w_i + c\right), \tag{11.15}$$

where b = output, f = activation function, a_i = input value, w_i = weight, and c = bias.

11.3.2 ANFIS

ANFISs are one of the effective time series supervised learning models that have been used for the optimization and prediction of various problems [36–38]. It can be used for efficient process control, task classification, problem-solving, and task forecasting. The model represents the input and output information of the engineered process, presenting it as "IF-THEN" fuzzy rules. It combines the learning strength of the fuzzy inference system (FIS) and the ANN, which deduces its membership function inputs from training examples. It is based on the first-order Sugeno fuzzy model, which utilizes either a backpropagation algorithm or a hybrid learning algorithm that aggregates the adaptive capacities of the neural network and the power of fuzzy logic [39]. The ANFIS model utilizes a neural network learning algorithm to generate a Takagi-Sugeno type of FIS that approximates non-linear systems using linear models. It is an adaptive, multilayer, and feedforward network that determines exact and uninterrupted functions [40]. It has five layers, where three layers consist of non-adaptive nodes and two adaptive nodes that assist the model in regulating a parameter based on the input and output datasets. The models are easy to develop compared to FIS because the rules are chosen based on the available data [41]. The set of rules with three (3) input variables b1, b2, and b3 and three Sugeno-type and fuzzy-Takagi if-then rules with one response (output) can be written as shown in Equations 11.16–11.18 [42].

Rule 1: If $(b_1 = C_1)$, $(b_2 = X_1)$, and $(b_3 = R_1)$ then

$$f_1 = n_1 b_1 + k_1 b_2 + m_1 b_3 + p_1. \tag{11.16}$$

Rule 2: If $(b_1 = c_2)$, $(b_2 = X_2)$, and $(b_3 = R_2)$, then

$$f_2 = n_2 b_1 + k_2 b_2 + m_2 b_3 + p_2. \tag{11.17}$$

Rule 3: If $(b_1 = c_3)$, $(b_2 = X_3)$, and $\left(b_3 = R_3\right)$, then

$$f_3 = n_3 b_1 + k_3 b_2 + m_3 b_3 + p_3, \tag{11.18}$$

where C, X *and* R are the fuzzy sets, and k, n, m *and* p are the nodal parameters.

Figure 11.3 presents the ANFIS architectural framework with two inputs, five fuzzy layers of normalization, product, defuzzification, and output layers, and can be presented as follows:

Adaptive nodes are in the first layer, known as the fuzzy layer, and the membership function determines the response. The output function can be estimated using Equation 11.19 [43].

$$H_{1,i} = \mu_{ci} B_i, \tag{11.19}$$

where μc_i is the membership function of node C.

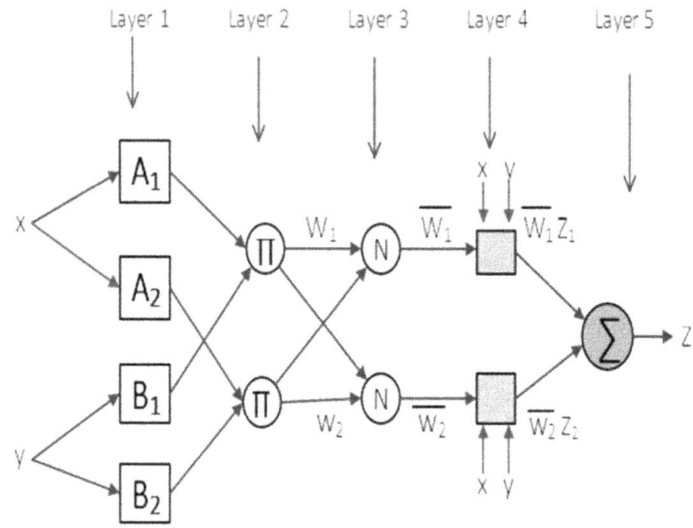

FIGURE 11.3 ANFIS architecture with two rules, two inputs, and one output [6].

The output of the nodes at the second layer defines the firing strength and output from the inputs. All the nodes at this layer are fixed and can be illustrated as shown in Equation 11.20.

$$H_{2,i} = w_i = \mu_{ci}\, b\, \mu_{Xi}\, b\, \mu_{Ri}\, b\, \mu_{pi}, \tag{11.20}$$

where μc_i, μ_{Xi}, μ_{Ri}, and μ_{pi} represent the membership functions at C, X, R, and P nodes, respectively, at the input layer.

At the third layer of the output is the quotient of the firing strength of the nodes to the sum of the cumulative capacities of the other nodes; this layer is called the normalized layer and can be represented by Equation 11.21.

$$H_{3,i} = \varpi_i = \frac{w_i}{\sum_j w_i}, \tag{11.21}$$

where w_i is the firing capacity at node i.

The fourth layer comprises the defuzzification layer, which utilizes the nodal function to recognize the impact of each rule at each node in the output directions. The summation of normalized signals from the initial nodes and the linear input signals yields the output, as illustrated in Equation 11.22.

$$H_{4,i} = H_{3,i}\, f_i = \varpi_i = w_i \left(n_i b + k_i y + m_i\right). \tag{11.22}$$

$\bar{w}_i$ is in the third layer, where the firing strength is normalized, f_i is the fuzzy rule, and n, k, and m are the parameters. The output layer is in the fifth layer with a single layer where all the signals from the previous layers are summed to produce the response and can be represented in Equation 11.23 [24].

$$H_{5,i} = \sum_i w_i f_i = \frac{\sum_i w_i f_i}{\sum_i w_i} = f_{out}. \tag{11.23}$$

During ANFIS modeling, there is a need for further data preparation before utilization to enhance performance and improve prediction accuracy [44]. This preparation includes normalization, handling missing data, denoising, outlier removal, filtering, and dimension reduction, among other steps. Data clustering is a crucial approach in data mining and statistics, which involves grouping datasets into distinct clusters and allocating them in a manner that the objects within a cluster differ from those in other clusters. Clustering methods are utilized in ANFIS to group the data into related fuzzy clusters, allocate membership functions, and produce the FIS arrangement from the data. The most used clustering methods in ANFIS include fuzzy c-means (FCM), grid partitioning (GP), and subtractive clustering (SC) [41].

The ANFIS model, in both its standalone and hybrid approaches, has been successfully applied in various biorefineries. For instance, the ANFIS model was applied to optimize and predict biogas yield from the anaerobic co-digestion of mango pulp, *Chromolaena odorata*, and cow dung, and it was reported to accurately fit the cumulative biogas yield [40]. In another biorefinery study, an ANFIS model was used to model and predict the cetane number of biodiesels based on the fatty acid methyl ester composition. The input parameters considered were the molar weight and the number of double bonds in biodiesel fuels, which were reported to have the potential to predict the cetane number [37]. Lignocellulose pyrolysis was predicted using an ANFIS model tuned with PSO, and the model successfully predicted the kinetic constants of biomass pyrolysis [36]. Figure 11.4 shows a plot of the ANFIS-predicted methane yield and the observed methane produced from the anaerobic digestion process [6]. It can be observed that the ANFIS predicted yield follows a similar path with the experimental yield, and the over- and under-prediction can be linked to the sensitivity of the model to marginal variation of the test data and response to unfamiliar input parameters of the digestion process, leading to the inaccurate forecasting at some points [45]. With the record of the ANFIS model's strength to predict bioenergy, there is a limited study on its application in biorefinery, and lignocellulose feedstocks are not given the required attention. Upon considering the input parameters, it was observed that one of the important input parameters of biorefinery, namely pretreatment, is currently missing from the literature. Therefore, further study is required to investigate the ability of the ANFIS model on different biorefinery processes when the input parameter of pretreatment is considered, which will support the upscaling and commercialization of the biorefinery process.

11.3.3 AI MODELS' PERFORMANCE METRICS

Investigating the performance and accuracy of AI models by comparing observed and predicted data can be done using various statistical indicators. Performance

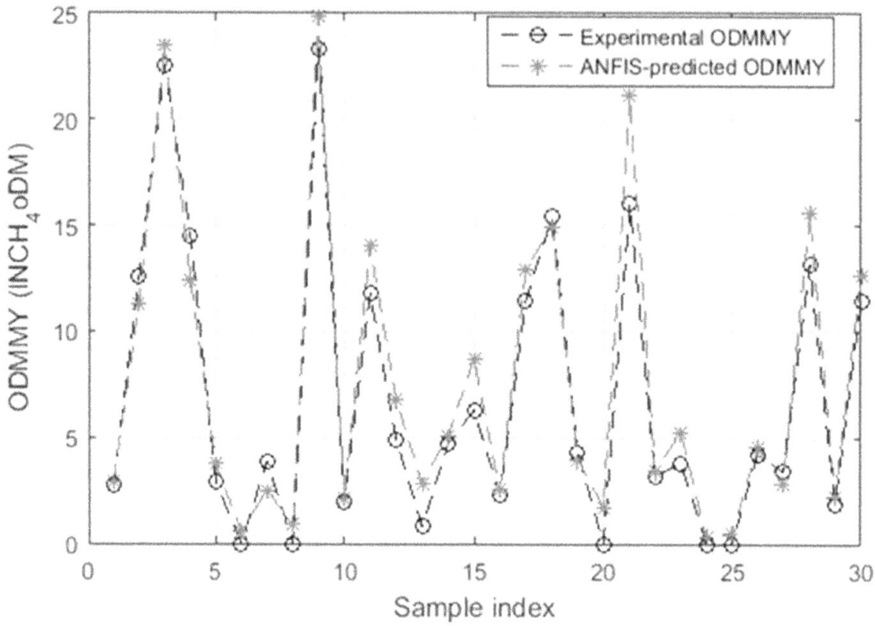

FIGURE 11.4 ANFIS predicted and experimental methane yield of lignocellulose biomass [6].

metrics can be used to assess the accuracy of the developed model in selecting the optimal model conditions. There are various parameters used in AI models, but the root mean square error (RMSE), mean absolute percentage error (MAPE), mean absolute deviation (MAD), and R-squared (R^2) are the most commonly used performance indicators [46]. RMSE and MAD are statistical tools used to assess the reliability and validity of the developed model in predicting bioenergy, with lower values indicating a better fit. MAPE determines the model accuracy, with lower values indicating a more accurate model. R2 values represent the relationship between the input data and the forecasted data, and values within the range of 0.75–1 (75–100%) indicate a suitable model for optimization and prediction techniques. RMSE, MAPE, MAD, and R^2 can be calculated using Equations 11.24–11.27 [47,48].

$$RMSE = \left(\sum_{i=1}^{N} \frac{(P_i - O_i)^2}{N} \right)^{1/2}, \tag{11.24}$$

$$MAPE = \frac{1}{N} \sum_{i=1}^{N} \left| \frac{O_i - P_i}{O_i} \right| x100\%, \tag{11.25}$$

$$MAD = \sum_{i=1}^{N} \frac{(O_i - P_i)}{N}, \tag{11.26}$$

$$R^2 = 1 - \left[\frac{\sum_{i=1}^{n} (O_i - O)(P_i - P)}{\sqrt{\sum_{i=1}^{n} (O_i - O)^2 \, x \sum_{i=1}^{n} (P_I - P)^2}} \right]^2, \tag{11.27}$$

where i = the sample index, O_i = the experimental bioenergy generated for the ith sample, O = the average experimental values, N = the number of samples, P_i = the predicted bioenergy yield for the ith sample, and P = the average predicted yields.

11.4 KINETIC MODELS

Bioprocessing approaches utilize cell components and microorganisms, such as enzymes, to produce new materials, including bioenergy, through the biorefining of readily available biomass. Enhancement of the biorefinery process requires understanding the relationship between independent input variables, such as biomass, temperature, substrate, pH, pretreatment, inhibitors, and light intensity, and dependent responses (output variables) [49]. The use of kinetic parameters and mathematical models to fit the observed kinetic data from laboratory pilot scales, using suitable equations such as linear or non-linear regression analysis, can enhance the level of understanding of the biorefinery process [50]. Mathematical modeling of the biorefinery process is a crucial stage in understanding biorefinery, as it reveals the parameters necessary for process optimization, simulation, and scaling up of biorefinery. Bioenergy biorefineries have been modeled using various kinetic models, such as logistic, first-order, Gompertz, Cone, Schnute, modified Gompertz, and Gompertz with second-order kinetics, among others [51–53]. Models selected should be able to adjust suitably to the kinetic profiles and accurately forecast the kinetic parameters, flow rate, output, and total bioenergy release while also generating metabolites and volatile fatty acids under varying operating conditions [54]. The use of simple and multiple linear regression in bioenergy production has been investigated widely. The benefit of using these variables in regression analysis is the inclusion of both the interaction between various inputs and non-linear data behavior, while the model remains linear and easy to interpret. Non-linear regression has been found to be the most suitable model for modeling the exponential growth of microorganisms [55]. Due to the need to unravel the knowledge on the performance of time-varying microbes and non-linear dynamics, progress has been made in the search for new techniques and designs to enhance mathematical modeling for the development and control of biorefineries. The biorefinery process involves various kinetics, including the growth kinetics of microbes, feedstock consumption, bioenergy production, and the impact of inhibitors on microorganism growth kinetics [56]. Kinetic models can assist in optimizing fermentation and pretreatment processes [57]. The kinetic models for bioenergy production can be categorized as discussed below:

11.4.1 MICROORGANISMS' GROWTH KINETIC MODELS

Microorganism growth involved lag, stationary, log, and death stages. At the lag stage, microbes adjust to the environment and set for proper growth, leading to minimal changes in population size. As the microbes developed to the log stage, the microorganism population rose exponentially until a peak growth rate was achieved. The process then transitions into a stationary stage, at which the microorganism's development rate equals its death rate [58]. Furthermore, at the death stage, the microorganism population falls as the development rate decreases below the death rate. During the entire growth stage, the lag time furnishes information on how fast microbes can adjust to the environment. The optimal microorganism size at the stationary stage facilitates the design of approaches for long-term management of microorganisms. This information can be generated using kinetic models, such as the first-order model (Equation 11.28), the Gompertz model (Equation 11.29), the Richards model (Equation 11.30), and the logistic model (Equation 11.31), during biorefinery [59]. The Monod (Equation 11.32) model has been widely utilized to analyze growth-linked feedstock utilization. It is useful mainly when a pseudo-steady state in the cells is attained and is not applicable to the lag phase [60]. Modified Monod (Equation 11.33) presents the impacts of initial feedstock concentration on the rate of growth [61].

$$N = N_o\left[1 - e\left(-et\right)\right], \tag{11.28}$$

$$N = Y.\exp\left[-\exp\left[\frac{\beta_m}{Y}(\lambda - t)\right] + 1\right], \tag{11.29}$$

$$N = A_o + \Delta A_{max}\left\{1 + cX\exp(1+c)X\exp\left[\frac{P_{max}}{\Delta A_{max}}(1+c)^{\left(1+\frac{1}{c}\right)}X(\lambda - \mathrm{t})\right]\right\}^{\left(-\frac{1}{c}\right)}, \tag{11.30}$$

$$N = \frac{N_o}{\left[1 + exp\left[\frac{4\beta_m}{Y}(\lambda - t) + 2\right]\right]}, \tag{11.31}$$

$$\mu = \mu_{max}X\frac{[B]}{R_s + [B]}, \tag{11.32}$$

$$\mu = \mu_{max}X\frac{[B]}{R_sS_o + [B]}, \tag{11.33}$$

where N is the total bioenergy generated during the retention period (ml/g), N_o is the specific bioenergy strength of the biomass (ml/g), e is the bioenergy generated rate constant (ml/day), t is the time (days), β_m = the highest bioenergy generated, A = the fractional weight, c = the mathematical constant, P_{max} = the maximum release rate of individual space (ml/ l.h), B = the biomass concentration, R_s= the half-saturation ($g.l^{-1}$), and λ = the lag phase (days).

11.4.2 Bioenergy Generation Kinetics

Bioenergy is released during the degradation of biomass, either through thermochemical or biological processes. In a biological process, microbes grow and consume biomass, resulting in the release of various products, including biomethane, biohydrogen, and bioethanol [62,63]. The ability of microbes to convert biomass into biofuel depends primarily on the concentration of biomass. It was observed that, within a convenient range, increasing biomass concentrations can effectively improve bioenergy release and biomass degradation; however, biomass concentrations at very high levels may lead to a decrease in these processes with rising levels [55]. During the biorefinery process, a lag phase exists, characterized by the initial slow growth of microbes and a delay between bioenergy generation and microbial growth. Furthermore, increasing microbial development results in an exponential increase in bioenergy release. However, once the death rate of microbes exceeds their growth rate, the microbial population begins to decline, resulting in a slowdown in the bioenergy release rate. Understanding the lag time, bioenergy release rate, and optimum bioenergy quantity is essential for optimizing bioenergy release. Bioenergy release kinetic models can be derived by incorporating the bioenergy release coefficient into microbial development kinetic models. The Ha-Levenspiel model, an extension of the Monod model presented in Equation 11.34, was developed to analyze the influence of biomass concentrations on bioenergy release rate. The model was noticed to fit the data better than the classical Monod model [64]. The modified Gompertz (Equation 11.35) and modified logistic (Equation 11.36) models have been widely used to fit bioenergy release data with respect to time, to analyze bioenergy release curves quantitatively, and to determine kinetic parameters. The modified Gompertz model was observed to predict bioenergy more accurately than the logistic model [65].

$$M = \frac{M_{o.max} A \left(1 - \frac{A}{A_m}\right)^c}{A + R_s}, \tag{11.36}$$

$$M = M_o . e \left\{ -e \left[\frac{A_m . e}{M_0} . (\lambda - t) + 1 \right] \right\}, \tag{11.35}$$

$$M = \frac{M_{max}}{\lambda + \frac{M_{max}}{A_m}} X \frac{1}{K}, \tag{11.36}$$

where M = total bioenergy released (ml), M_0 = specific bioenergy release of feedstock (ml), A_m = optimum specific bioenergy release rate (ml /day), K = working volume of the culture (ml or m^3), λ = lag phase, and t = time (days).

11.4.3 Biomass consumption kinetics

Microbes use biomass as a nutrient source and energy for different metabolic functions. Understanding the kinetics related to biomass consumption, including the relationship between microorganism development rate and biomass concentration, bioenergy generation and biomass concentration, and the variation in biomass concentration over time, is crucial for identifying optimal biomass conditions to enhance bioenergy production and mitigate inhibitory impacts. The Monod kinetic model was utilized as an empirical cell development model and further used in forecasting biomass consumption kinetics [66]. It can be used to determine the optimal growth rate and biomass concentration for half of the optimal development rate. This model has been widely used to analyze biomass concentration kinetics during bioenergy production, enabling the determination of optimum generation rate. It can also forecast the changes in biomass concentration with time. It is essential to note that the Monod model shares a feature with the Michaelis–Menten model (Equation 11.37), which is often misinterpreted in its application [67]. Nonetheless, these models need to be differentiated since their deductions, input variables, and physical meanings vary. The traditional Monod model cannot sufficiently fit bioenergy production when the higher concentration of biomass inhibits it. Studies have shown that the Monod equation cannot adequately fit the bioenergy production from lignocellulose biomass due to the inability of cellulose to degrade completely [68,69]. Therefore, Andrew's model, illustrated in Equation 11.38, was specifically developed to analyze the growth behavior of microbes with a biomass inhibition term. This model provides information about optimum growth behavior, saturation, and inhibition constants [70]. Growth kinetics, such as Gompertz and logistic models, can be combined with the generation coefficient to analyze biomass consumption kinetic curves.

$$M = \frac{uz_m}{h_m + z_m}, \tag{11.37}$$

$$Y = \frac{Y_{max}}{1 + \frac{c_s}{a} + \frac{a}{c_i}}, \tag{11.38}$$

where M = reaction rate, u = maximum reaction rate, z_m = biomass concentration, hm = biomass concentration where the rate is $\frac{1}{2}$ M, Y = optimum consumption rate, c = saturation constant, and a = optimum biomass concentration.

11.5 CONCLUSION

Mathematical modeling of the biorefinery process aids in interpreting bioenergy production and provides clear insights into the response of biomass to environmental and other process parameters. It also predicts the corresponding kinetic constants and reveals important information on the biorefinery input parameters that can be used to optimize and predict bioenergy production from biomass. Various statistical, AI, and kinetic mathematical models have been applied to optimize and predict bioenergy. The performance of these models varies across all biorefinery processes due to differences in process parameters among microbes that generate bioenergy through their metabolism, anatomy, response, and physiology. The uniformity of the models may likely be the same if the bioprocesses are the same. Despite the successes recorded in the application of mathematical modeling in biorefinery, it was noted that there is a limited study that considers pretreatment application as an input parameter. Considering the importance of pretreatment application in the successful biorefinery process, there is an urgent need for more studies that focus on modeling the biorefinery process when pretreatment is considered. This will facilitate the upscaling of various biorefinery processes, save time, and promote industrial production. Selection of the most efficient models under varying conditions does not have a standard yet. This limitation of clarity in model selection hinders the application of mathematical models; therefore, there is a need for unified guidelines that assist in selecting suitable models across various biorefinery process conditions.

REFERENCES

[1] A. Tshikovhi, T.E. Motaung, Technologies and Innovations for Biomass Energy Production, Sustain 15 (2023) 12121. https://doi.org/10.3390/SU151612121.
[2] R.N. Ossei-Bremang, F. Kemausuor, A decision support system for the selection of sustainable biomass resources for bioenergy production, Environ Syst Decis41 (2021) 437–454. https://doi.org/10.1007/S10669-021-09810-6.
[3] O. Adeleke, K.O. Olatunji, D.M. Madyira, T.C. Jen, Application of multimodal machine learning-based analysis for the biomethane yields of NaOH-pretreated biomass, Sci Rep 15 (2025) 1–20. https://doi.org/10.1038/s41598-025-09527-5.
[4] A. Welfle, P. Thornley, M. Röder, A review of the role of bioenergy modelling in renewable energy research & policy development, Biomass Bioenergy 136 (2020) 105542. https://doi.org/10.1016/J.BIOMBIOE.2020.105542.
[5] K.O. Olatunji, O. Adeleke, T.C. Jen, D.M. Madyira, Impact of acidic pretreatment on biomethane yield from Xyris capensis: Experimental and in-depth data-driven insight, Processes 13 (2025). https://doi.org/10.3390/PR13071997.
[6] K.O. Olatunji, N.A. Ahmed, D.M. Madyira, A.O. Adebayo, O. Ogunkunle, O. Adeleke, Performance evaluation of ANFIS and RSM modeling in predicting biogas and methane yields from Arachis hypogea shells pretreated with size reduction, Renew Energy 189 (2022) 288–303. https://doi.org/10.1016/J.RENENE.2022.02.088.
[7] C. Escamilla-Alvarado, E. Ríos-Leal, M.T. Ponce-Noyola, H.M. Poggi-Varaldo, Gas biofuels from solid substrate hydrogenogenic-methanogenic fermentation of the organic fraction of municipal solid waste, Process Biochemistry, Elsevier (2012) 1572–1587. https://doi.org/10.1016/j.procbio.2011.12.006.
[8] P. Noll, M. Henkel, History and evolution of modeling in biotechnology: Modeling & simulation, application and hardware performance, Comput Struct Biotechnol J 18 (2020) 3309. https://doi.org/10.1016/J.CSBJ.2020.10.018.
[9] C.P. Mitchell, Development of decision support systems for bioenergy applications, Biomass Bioenergy 18 (2000) 265–278. https://doi.org/10.1016/S0961-9534(99)00099-9.
[10] K.O. Olatunji, D.M. Madyira, N.A. Ahmed, S.O. Jekayinfa, O. Ogunkunle, Modelling the effects of particle size pretreatment method on biogas yield of groundnut shells, Waste Manag Res 40 (2022). https://doi.org/10.1177/0734242X211073852.
[11] Ç. Gönen, E.Ü. Deveci, N. Akter Önal, Evaluation of biomass pretreatment to optimize process factors for different organic acids via Box–Behnken RSM method, J Mater Cycles Waste Manag 23 (2021) 2016–2027. https://doi.org/10.1007/S10163-021-01276-7/TABLES/8.
[12] L. Su, X. Sun, C. Liu, R. Ji, G. Zhen, M. Chen, L. Zhang, Thermophilic solid-state anaerobic digestion of corn straw, cattle manure, and vegetable waste: Effect of temperature, total solid content, and C/N ratio, Archaea 2020 (2020). https://doi.org/10.1155/2020/8841490.
[13] I. Arslan-Alaton Idil, A.B. Yalabik, T. Olmez-Hanci, Development of experimental design models to predict Photo-Fenton oxidation of a commercially important naphthalene sulfonate and its organic carbon content, Chem Eng J 165 (2010) 597–606. https://doi.org/10.1016/j.cej.2010.10.003.

[14] S. Sathish, S. Vivekanandan, Parametric optimization for floating drum anaerobic bio-digester using response surface methodology and artificial neural network, Alex Eng J 55 (2016) 3297–3307. https://doi.org/10.1016/J.AEJ.2016.08.010.

[15] K.O. Olatunji, D.M. Madyira, Response surface methodology, multi-objective optimization of biogas, and methane yield of pretreated Arachis hypogea shells, Mater Today Proc 105 (2024) 85–91. https://doi.org/10.1016/J.MATPR.2023.07.351.

[16] K.O. Olatunji, D.M. Madyira, N.A. Ahmed, O. Adeleke, O. Ogunkunle, Biomethane yield modeling and optimization from thermally pretreated Arachis hypogea shells using response surface methodology and artificial neural network, Bioresour Technol Rep 20 (2022). https://doi.org/10.1016/j.biteb.2022.101236.

[17] K.O. Olatunji, D.M. Madyira, Application of response surface methodology (RSM) for optimizing methane yield of oxidative pretreated Xyris capensis, in: E3S Web of Conferences 433 (2023) 01007. https://doi.org/10.1051/E3SCONF/202343301007.

[18] E. Betiku, S.O. Ajala, Modeling and optimization of Thevetia peruviana (yellow oleander) oil biodiesel synthesis via Musa paradisiacal (plantain) peels as heterogeneous base catalyst: A case of artificial neural network vs. response surface methodology, Ind Crops Prod 53 (2014) 314–322. https://doi.org/10.1016/J.INDCROP.2013.12.046.

[19] A.M. Akintunde, S.O. Ajala, E. Betiku, Optimization of bauhinia monandra seed oil extraction via artificial neural network and response surface methodology: A potential biofuel candidate, Ind Crops Prod 67 (2015) 387–394. https://doi.org/10.1016/J.INDCROP.2015.01.056.

[20] K.O. Olatunji, D.M. Madyira, Effect of acidic pretreatment on the microstructural arrangement and anaerobic digestion of Arachis hypogea shells; and process parameters optimization using response surface methodology, Heliyon 9 (2023) e15145. https://doi.org/10.1016/J.HELIYON.2023.E15145.

[21] Y. Sewsynker-Sukai, F. Faloye, G. Kana, Artificial neural networks: An efficient tool for modelling and optimization of biofuel production (a mini review), Biotechnol Biotechnol Equip 31 (2016). https://doi.org/10.1080/13102818.2016.1269616.

[22] B. Najafi, S. Faizollahzadeh Ardabili, Application of ANFIS, ANN, and logistic methods in estimating biogas production from spent mushroom compost (SMC), Resour Conserv Recycl 133 (2018) 169–178. https://doi.org/10.1016/j.resconrec.2018.02.025.

[23] I. Malekmohamadi, M.R. Bazargan-Lari, R. Kerachian, M.R. Nikoo, M. Fallahnia, Evaluating the efficacy of SVMs, BNs, ANNs and ANFIS in wave height prediction, Ocean Eng 38 (2011) 487–497. https://doi.org/10.1016/j.oceaneng.2010.11.020.

[24] K.O. Olatunji, S.O. Oladipo, D.M. Madyira, Y. Sun, Performance evaluation of different clustering techniques and parameters of hybrid PSO- and GA-ANFIS on optimization and prediction of biomethane yield of alkali-pretreated groundnut shells, Waste Biomass Valor 16 (2024) 423–440. https://doi.org/10.1007/S12649-024-02674-2/TABLES/10.

[25] K. Ukoba, O.R. Onisuru, T.-C. Jen, D.M. Madyira, K.O. Olatunji, Predictive modeling of climate change impacts using Artificial Intelligence: A review for equitable governance and sustainable outcome, Environ Sci Pollut Res 2025 (2025) 1–20. https://doi.org/10.1007/S11356-025-36356-W.

[26] O. Olatunji, S. Akinlabi, N. Madushele, P. Adedeji, S. Fatoba, Evolution algorithms and biomass properties prediction: A review, in: American Society of Mechanical Engineers, Power Division (Publication) POWER, American Society of Mechanical Engineers (ASME) (2019). https://doi.org/10.1115/POWER2019-1826.

[27] Y. Sewsynker-Sukai, F. Faloye, G. Kana, Artificial neural networks: An efficient tool for modelling and optimization of biofuel production (a mini review), Biotechnol Biotechnol Equip 31 (2016). https://doi.org/10.1080/13102818.2016.1269616.

[28] O. Adeleke, S.A. Akinlabi, T.-C. Jen, I. Dunmade, Application of artificial neural networks for predicting the physical composition of municipal solid waste: An assessment of the impact of seasonal variation, Waste Manag Res (2021) 734242X21991642. https://doi.org/10.1177/0734242X21991642.

[29] F. Almomani, Prediction of biogas production from chemically treated co-digested agricultural waste using artificial neural network, Fuel 280 (2020) 118573. https://doi.org/10.1016/J.FUEL.2020.118573.

[30] K.O. Olatunji, D.M. Madyira, N.A. Ahmed, O. Adeleke, O. Ogunkunle, Modeling the biogas and methane yield from anaerobic digestion of Arachis hypogea shells with combined pretreatment techniques using machine learning approaches, Waste Biomass Valor 14 (2023). https://doi.org/10.1007/s12649-022-01935-2.

[31] N.D. Nguyen, V.T. Nguyen, Development of ANN structural optimization framework for data-driven prediction of local two-phase flow parameters, Prog Nucl Energy 146 (2022) 104176. https://doi.org/10.1016/j.pnucene.2022.104176.

[32] F. Almomani, Prediction of biogas production from chemically treated co-digested agricultural waste using artificial neural network, Fuel 280 (2020) 118573. https://doi.org/10.1016/J.FUEL.2020.118573.

[33] J.M. Avramović, A.V. Veličković, O.S. Stamenković, K.M. Rajković, P.S. Milić, V.B. Veljković, Optimization of sunflower oil ethanolysis catalyzed by calcium oxide: RSM versus ANN-GA, Energy Convers Manag 105 (2015) 1149–1156. https://doi.org/10.1016/J.ENCONMAN.2015.08.072.

[34] C.N. Igwilo, N.C. Ude, I.M. Onoh, C.B. Enekwe, B.A. Alieze, RSM, ANN and ANFIS applications in modeling fermentable sugar production from enzymatic hydrolysis of Colocynthis Vulgaris Shrad seeds shell, Bioresour Technol Rep 18 (2022) 101056. https://doi.org/10.1016/J.BITEB.2022.101056.

[35] E.G. Aklilu, Y.A. Waday, Optimizing the process parameters to maximize biogas yield from anaerobic co-digestion of alkali-treated corn stover and poultry manure using artificial neural network and response surface methodology, Biomass Convers Biorefin 2021 (2021) 1–14. https://doi.org/10.1007/S13399-021-01966-0.

[36] M. Aghbashlo, M. Tabatabaei, M.H. Nadian, V. Davoodnia, S. Soltanian, Prognostication of lignocellulosic biomass pyrolysis behavior using ANFIS model tuned by PSO algorithm, Fuel 253 (2019) 189–198. https://doi.org/10.1016/J.FUEL.2019.04.169.

[37] M. Mostafaei, ANFIS models for prediction of biodiesel fuels cetane number using desirability function, Fuel 216 (2018) 665–672. https://doi.org/10.1016/J.FUEL.2017.12.025.

[38] K.O. Olatunji, D.M. Madyira, O. Adeleke, Optimizing anaerobic co-digestion of Xyris capensis and duck waste using neuro-fuzzy model and response surface methodology, Fuel 354 (2023) 129334. https://doi.org/10.1016/J.FUEL.2023.129334.

[39] P.A. Adedeji, S.O. Masebinu, S.A. Akinlabi, N. Madushele, Adaptive Neuro-fuzzy Inference System (ANFIS) modelling in energy system and water resources, Optimization Using Evolutionary Algorithms and Metaheuristics, CRC Press (2019) 117–133. https://doi.org/10.1201/9780429293030-7.

[40] M.O. Fajobi, O.A. Lasode, A.A. Adeleke, P.P. Ikubanni, A.O. Balogun, P. Paramasivam, Prediction of biogas yield from codigestion of lignocellulosic biomass using Adaptive Neuro-Fuzzy Inference System (ANFIS) Model, J Eng 2023 (2023) 9335814. https://doi.org/10.1155/2023/9335814.

[41] O. Adeleke, S.A. Akinlabi, T.C. Jen, I. Dunmade, Prediction of municipal solid waste generation: An investigation of the effect of clustering techniques and parameters on ANFIS model performance, Environ Technol (United Kingdom) (2020). https://doi.org/10.1080/09593330.2020.1845819.

[42] A. Azad, M. Manoochehri, H. Kashi, S. Farzin, H. Karami, V. Nourani, J. Shiri, Comparative evaluation of intelligent algorithms to improve adaptive neuro-fuzzy inference system performance in precipitation modelling, J Hydrol (Amst) 571 (2019) 214–224. https://doi.org/10.1016/j.jhydrol.2019.01.062.
[43] J.-S. Roger Jang, ANFIS : Adaptive-Network-Based Fuzzy Inference System, IEEE Trans Syst Man Cybern 23 (1993).
[44] I. Okumus, A. Dinler, Current status of wind energy forecasting and a hybrid method for hourly predictions, Energy Convers Manag 123 (2016) 362–371. https://doi.org/10.1016/j.enconman.2016.06.053.
[45] I. Okumus, A. Dinler, Current status of wind energy forecasting and a hybrid method for hourly predictions, Energy Convers Manag 123 (2016) 362–371. https://doi.org/10.1016/j.enconman.2016.06.053.
[46] P.A. Adedeji, S. Akinlabi, N. Madushele, O.O. Olatunji, Wind turbine power output very short-term forecast: A comparative study of data clustering techniques in a PSO-ANFIS model, J Clean Prod 254 (2020) 120135. https://doi.org/10.1016/j.jclepro.2020.120135.
[47] P.A. Adedeji, S.O. Masebinu, S.A. Akinlabi, N. Madushele, Adaptive Neuro-fuzzy Inference System (ANFIS) modelling in energy system and water resources, Optimization Using Evolutionary Algorithms and Metaheuristics, CRC Press (2019) 117–133. https://doi.org/10.1201/9780429293030-7.
[48] K.O. Olatunji, D.M. Madyira, Influence of different clustering techniques of ANFIS on biomethane yield optimization and prediction from eutectic solvent pretreated groundnut shells, in: IEEE International Conference on Emerging and Sustainable Technologies for Power and ICT in a Developing Society, NIGERCON (2024). https://doi.org/10.1109/NIGERCON62786.2024.10927103.
[49] S.K. Pramanik, F.B. Suja, M. Porhemmat, B.K. Pramanik, Performance and kinetic model of a single-stage anaerobic digestion system operated at different successive operating stages for the treatment of food waste, Processes 7 (2019) 600. https://doi.org/10.3390/PR7090600.
[50] Y. Pererva, C.D. Miller, R.C. Sims, Existing empirical kinetic models in Biochemical Methane Potential (BMP) testing, their selection and numerical solution, Water 12 (2020) 1831. https://doi.org/10.3390/W12061831.
[51] R. Karki, W. Chuenchart, K.C. Surendra, S. Sung, L. Raskin, S.K. Khanal, Anaerobic co-digestion of various organic wastes: Kinetic modeling and synergistic impact evaluation, Bioresour Technol 343 (2022) 126063. https://doi.org/10.1016/J.BIORTECH.2021.126063.
[52] K.O. Olatunji, D.M. Madyira, Performance evaluation of three kinetic models on biomethane yield of nano-additive pretreated Xyris capensis, in: IEEE International Conference on Emerging and Sustainable Technologies for Power and ICT in a Developing Society, NIGERCON (2024). https://doi.org/10.1109/NIGERCON62786.2024.10927124.
[53] K.O. Olatunji, A.D. Olugbemide, R.F. Akerejola, D.M. Madyira, Kinetic modelling of the biomethane production potential of acidic pretreated groundnut shells, Biomass Convers Biorefin (2025) 1–15. https://doi.org/10.1007/S13399-025-06549-X/FIGURES/5.
[54] H. Akroum, D. Akroum-Amrouche, A. Aibeche, H. Akroum, D. Akroum-Amrouche, A. Aibeche, Modeling methods used in bioenergy production processes: A review, Adv Comput Des 5 (2020) 323. https://doi.org/10.12989/ACD.2020.5.3.323.
[55] H. Akroum, D. Akroum-Amrouche, A. Aibeche, Modeling of conversion kinetics in bioenergy production technologies, Int J Green Energy 21 (2024) 3415–3430. https://doi.org/10.1080/15435075.2024.2377355.

[56] X. Guo, J. Wang, Guidelines for selection and application of kinetics models in bioproduction processes, Trends Biotechnol 43 (2025) 1302–1316. https://doi.org/10.1016/J.TIBTECH.2024.11.007.

[57] K.O. Olatunji, A.D. Olugbemide, R.F. Akerejola, D.M. Madyira, Application of three kinetic models for the prediction of biomethane yield of combined oxidative and nanoparticle additives pretreated Xyris capensis, Bioenergy Res 18 (2025) 86. https://doi.org/10.1007/S12155-025-10891-3.

[58] R.L. Buchanan, R.C. Whiting, W.C. Damert, When is simple good enough: A comparison of the Gompertz, Baranyi, and three-phase linear models for fitting bacterial growth curves, Food Microbiol 14 (1997) 313–326. https://doi.org/10.1006/FMIC.1997.0125.

[59] A.M. Abubakar, K. Silas, M.M. Aji, U.H. Taura, J. Undiandeye, Microbial growth rate kinetics in biogas production by anaerobic digestion of chicken manure, IRASD J Energy Environ 3 (2022) 72–89. https://doi.org/10.52131/JEE.2022.0302.0027.

[60] O. Mohiuddin, A.P. Harvey, S. Tabraiz, M.T. Ameen, S. Velasquez-Orta, Kinetic modelling of yeast growth and pollutant removal in secondary effluent, J Water Process Eng 50 (2022) 103244. https://doi.org/10.1016/J.JWPE.2022.103244.

[61] F. Ahmad, A.T. Jameel, M.H. Kamarudin, M. Mel, Study of growth kinetic and modeling of ethanol production by Saccharomyces cerevisae, Afr J Biotechnol 10 (2011) 18842–18846. https://doi.org/10.5897/AJB11.2763.

[62] T. Lepage, M. Kammoun, Q. Schmetz, A. Richel, Biomass-to-hydrogen: A review of main routes production, processes evaluation and techno-economical assessment, Biomass Bioenergy 144 (2021) 105920. https://doi.org/10.1016/J.BIOMBIOE.2020.105920.

[63] F. Dalena, A. Senatore, A. Iulianelli, L. Di Paola, M. Basile, A. Basile, Ethanol from biomass: Future and perspectives, Ethanol: Science and Engineering (2019) 25–59. https://doi.org/10.1016/B978-0-12-811458-2.00002-X.

[64] M.R. Boni, S. Sbaffoni, L. Tuccinardi, P. Viotti, Development and calibration of a model for biohydrogen production from organic waste, Waste Management 33 (2013) 1128–1135. https://doi.org/10.1016/J.WASMAN.2013.01.019.

[65] K. Jaman, N. Amir, M.A. Musa, A. Zainal, L. Yahya, A.M.A. Wahab, S. Suhartini, T.N.T.M. Marzuki, R. Harun, S. Idrus, Anaerobic digestion, codigestion of food waste, and chicken dung: Correlation of kinetic parameters with digester performance and on-farm electrical energy generation potential, Fermentation 8 (2022) 28. https://doi.org/10.3390/FERMENTATION8010028.

[66] J. Monod, The growth of bacterial cultures, Annu Rev Microbiol 3 (1949) 371–394. https://doi.org/10.1146/ANNUREV.MI.03.100149.002103.

[67] J. Wang, X. Guo, The Gompertz model and its applications in microbial growth and bioproduction kinetics: Past, present and future, Biotechnol Adv 72 (2024) 108335. https://doi.org/10.1016/J.BIOTECHADV.2024.108335.

[68] X. Yuan, X. Shi, P. Zhang, Y. Wei, R. Guo, L. Wang, Anaerobic biohydrogen production from wheat stalk by mixed microflora: Kinetic model and particle size influence, Bioresour Technol 102 (2011) 9007–9012. https://doi.org/10.1016/J.BIORTECH.2011.06.045.

[69] E. Mardawatia, D.W. Wira, M. Djali, F. Fetriyuna, E. Suryadi, Optimization and kinetic modelling of the enzymatic hydrolysis of oil palm petioles, KnE Life Sci 2 (2017) 439–447. https://doi.org/10.18502/KLS.V2I6.1065.

[70] J.F. Andrews, A mathematical model for the continuous culture of microorganisms utilizing inhibitory substrates, Biotechnol Bioeng 10 (1968) 707–723. https://doi.org/10.1002/BIT.260100602.

Index

For Product Safety Concerns and Information please contact our EU
representative GPSR@taylorandfrancis.com
Taylor & Francis Verlag GmbH, Kaufingerstraße 24, 80331 München, Germany

www.ingramcontent.com/pod-product-compliance
Ingram Content Group UK Ltd.
Pitfield, Milton Keynes, MK11 3LW, UK
UKHW020201200926
475702UK00004B/558
* 9 7 8 1 0 4 1 0 0 4 6 1 5 *